National Insecurity
U.S. Intelligence After the Cold War

National Insecurity

U.S. Intelligence After the Cold War

Edited by
CRAIG EISENDRATH

Foreword by
SENATOR TOM HARKIN

A Project of the Center for International Policy

TEMPLE UNIVERSITY PRESS
Philadelphia

Temple University Press, Philadelphia 19122

Copyright © 2000 by Temple University
All rights reserved
Published 2000
Printed in the United States of America

♾ The paper used in this publication meets the requirements of the American National Standard for Information Sciences—Permanence of Paper for Printed Library Materials, ANSI Z39.48-1984

Library of Congress Cataloging-in-Publication Data

National insecurity : U.S. intelligence after the Cold War / edited by
  Craig Eisendrath.
    p. cm.
  "A project of the Center for International Policy."
  Includes bibliographical references and index.
  ISBN 1-56639-744-8 (cloth : alk. paper)
  ISBN 1-56639-848-7 (paper)
  1. Intelligence service—United States. 2. World politics—1989–
I. Eisendrath, Craig R. II. Center for International Policy
(Washington, D.C.)
UB251.U5N38   1999
327.1273'09045—DC21   99-23807
                                                        CIP

Contents

Foreword
SENATOR TOM HARKIN ... vii

Introduction
CRAIG EISENDRATH ... 1

1 After the Cold War: The Need for Intelligence
ROGER HILSMAN ... 8

2 Espionage and Covert Action
MELVIN A. GOODMAN ... 23

3 Too Many Spies, Too Little Intelligence
ROBERT E. WHITE ... 45

4 CIA–Foreign Service Relations
ROBERT V. KEELEY ... 61

5 Covert Operations: The Blowback Problem
JACK A. BLUM ... 76

6 The End of Secrecy: U.S. National Security and the New Openness Movement
KATE DOYLE ... 92

7 Mission Myopia: Narcotics as Fallout From the CIA's Covert Wars
ALFRED W. MCCOY ... 118

8 TECHINT: The NSA, the NRO, and NIMA
ROBERT DREYFUSS ... 149

9	Improving the Output of Intelligence: Priorities, Managerial Changes, and Funding RICHARD A. STUBBING	172
10	Who's Watching the Store? Executive-Branch and Congressional Surveillance PAT M. HOLT	190

Conclusions
CRAIG EISENDRATH — 212

Selected Bibliography — 223

About The Center for International Policy — 227

About the Contributors — 231

Index — 233

SENATOR TOM HARKIN

Foreword

THE COLD WAR passed into history in 1989 with the demise of the Soviet Union. The world is still a treacherous place, but there is no longer a single hostile nation representing the ultimate threat to our country.

Instead, the U.S. military and intelligence institutions now focus on a host of foreign nations and emerging threats. Rogue states and terrorists are capable of inflicting great damage. Caches of old weapons can fall into the wrong hands. Global threats such as environmental disasters and large-scale refugee problems dominate the news.

The United States has the option of continuing on the path we embarked upon in the 1940s. We were a nation that, when not at war, took every international opportunity to seize the advantage in the struggle against the Soviet Union. We can continue to engage in covert actions. We can continue to conduct disinformation campaigns and to court military despots. And we can continue to conduct espionage on friends, foes, and neutrals. We can maintain vast military and intelligence structures geared up for a final showdown.

Such a stance no longer furthers security, however. Instead, it fosters an insecure world. Our current intelligence establishment is designed for the Cold War. The mandate of this institution should be reevaluated against the backdrop of the post–Cold War world, and we must make changes to keep up with a vastly different international situation.

To be sure, intelligence is still required, but it should be refocused. Its tactics should be circumscribed by the need to build a safer world, based on law and cooperation. Its operations should be less secret and more integrated with the needs of an open and dynamic foreign policy. It is time to forge a new path.

Twenty-five years ago, when Congressman Donald Fraser and I started the Center for International Policy, our nation's support for dictators and our questionable tactics abroad had earned us a dubious reputation around the world as hostile to human rights. Congressman Fraser and I believed that we needed to prepare the ground for a foreign policy that would more closely reflect our nation's democratic principles.

For decades, I have fought for a foreign policy based on human rights and the promotion of democratic ideals. I believe reform of our intelligence system is a necessity if we are to achieve that goal.

This book, a project of the Center for International Policy, written by a body of distinguished contributors, is an important step in this direction. It envisions an intelligence system with limited targets—a system that weighs with extreme care the use of covert action and espionage. It makes clear that such tactics should be employed only when our national interest is directly involved, when no other means are available, and, even then, only after determining that the dangers to be averted far outweigh any negative consequences of the action.

This book points to our diplomatic service as the primary government vehicle of foreign affairs. It asks that the use of secrecy in our intelligence operations be commensurate with national-security requirements. The need for secrecy must be legitimate. National security may no longer be used as an official excuse for avoiding government oversight or public debate.

We need a more rigorous system of supervision by both Congress and the executive branch if we are to avoid the abuses of previous decades. Cold War budgets are no longer justifiable; nor should they remain secret. We need a leaner, smaller intelligence establishment. Deployment of intelligence resources must be subject to standards of cost-effectiveness and democratic fairness. Congress owes it to all Americans to subject our intelligence system to the same scrutiny and standards it uses to evaluate domestic programs.

The distinguished contributors to this book present a wide range of perspectives from which to assess our intelligence system. Their decades of public service command tremendous respect. Their views break new ground and demand the attention of the White House and of lawmakers on both sides of the aisle. This book should be required reading by all congressional committees concerned with intelligence policy, surveillance, and appropriations, and by all Americans.

National Insecurity
U.S. Intelligence After the Cold War

CRAIG EISENDRATH

Introduction

THE UNITED STATES intelligence system is badly in need of reform. Its budget—$29 billion in 1998—and its mode of operation still reflect the life-or-death view of international relations of the Cold War. Between 1949 and the fall of the Berlin Wall, no cost was deemed too high and no deed too brutal when survival seemed at stake. In no area was this truer than in the areas of U.S. intelligence responsible for espionage and covert action. Paramilitary operations, election rigging, disinformation, massive electronic eavesdropping, and common cause with a host of the world's most undesirable characters all seemed justified, despite the fact that these actions systematically eroded our reputation around the world and our credibility at home.

We now know that our intelligence system served us badly even during the Cold War by providing unreliable intelligence on the Soviet Union and other areas of Cold War conflict, and by involving itself in repeated scandals. It succumbed to political pressure to "cook the books." By radically overestimating Soviet military strength and economic capacity, U.S. intelligence goaded this country to vastly increase its military budget, resulting in massive national debt and decreased standards of living for most Americans. Perhaps even worse, U.S. intelligence wreaked havoc around the world, supporting dictatorships, promoting the international narcotics trade, and repeatedly violating both foreign and domestic laws. Yet, protected by a shroud of secrecy, the intelligence establishment paid no price for its mistakes; year by year, it grew larger and more insulated from criticism. What Allen Dulles, former director of the CIA, once called "the craft of intelligence" became an ingrown cult of negligence and opportunism.

In October 1998, almost ten years after the end of the Cold War, the United States intelligence community received what Tim Weiner of the *New York Times* called "the largest spending increase for intelligence in 15 years." The increase came despite a long string of intelligence failures, including the intelligence community's inability "to foresee India's nuclear tests, to quickly and accurately assess a North Korean missile launch, to come up with a workable plan to undermine the Iraqi leader, Saddam Hussein, and to provide sharp analysis of global financial crises."[1]

Rather than justifying a budget cut, such failings, according to George J. Tenet, Director of Central Intelligence, made the case that the CIA, and implicitly the intelligence community as a whole, should receive more funds to create a greater and more reliable intelligence capacity. According to Tenet, the CIA had to "rebuild our field strength" in order "to stay in the espionage business." These views were seconded by Representative Porter J. Goss of Florida, chairman of the House Intelligence Committee. "This is going to be a long-term rebuild," declared Goss, a former CIA operations officer.

The writers of this book believe that increasing the budget of the intelligence community and expanding its role[2] are steps in the wrong direction. There clearly are important roles for an intelligence system in the post–Cold War era. The proliferation of nuclear, chemical, and bacteriological weapons and the growth of international terrorism continue to pose threats to the United States. There are threatening states whose foreign relations, operating in secrecy, may not yield adequate information through open sources. There is also a continuing need to centralize intelligence analysis coming in from a variety of sources, overt and covert, to provide U.S. policymakers with cogent analysis.

But in a world that does not currently pose serious strategic threats to the United States, is there still justification for the espionage and covert action that have been the stock-in-trade of our intelligence system during its first fifty years? Do we still need an intelligence system of close to the same size? As a *New York Times* editorial, commenting on the intelligence budget of $29 billion, put it, "Since well over two-thirds of the Cold War budget went into collecting intelligence about the Soviet Union, it is hard to understand why so much money is needed today, even to combat terrorism, develop new satellites and hire new staff."[3]

The authors of this book argue that we need to take a hard look at the entire U.S. intelligence establishment and make systematic reforms, including cutbacks rather than increases in its operations and budget. We should be clearheaded about what we need intelligence for, particularly intelligence secured through clandestine means, and to what extent covert action is in the interests of the United States. We need to weigh the enormous price of U.S. intelligence operations, not only in terms of dollars but in terms of the human suffering they cause abroad and at home. We also must weigh their effects on the credibility of our system of government. If we are considered bullies rather than friends by the peoples of many nations of the world, it is in no small part the fault of our intelligence system, and it can be corrected.

By 1994, a major revision of our intelligence operations was already long overdue. Referring to the CIA, Senator Arlen Specter of Pennsylvania, who was soon to become chairman of the Senate Select Committee on Intelligence, declared, "The place just needs a total overhaul." His dramatic statement was made at the time that a commission of House, Senate, and executive-branch members was created to review and possibly reform U.S. intelligence. As Senator Specter noted, "We are spending a lot of money on the CIA and there have been doubts for years as to whether we are getting our money's worth."[4]

These doubts about the intelligence system, shared by millions of Americans and many of Senator Specter's colleagues, sparked two major governmental studies. One was conducted by a joint executive-branch and congressional commission, headed by Harold Brown, former Secretary of Defense. The other was undertaken by the House Intelligence Committee; Representative Larry Combest, a Texas Republican, was the chairman. Their reports, issued in 1996,[5] came on the heels of the headline-making intelligence failures detailed in this book, which stretched from Guatemala to Iraq. Despite this dismal record, the 1996 reports did not call for basic reform, and there has been no call within the government for basic reform since they were issued.

Evidence of scandal, failure, and obfuscation has continued to mount. A June 1996 report by the Intelligence Oversight Board stated that the CIA had knowingly supported Guatemalan military officers implicated in political assassination, extrajudicial killing, kidnapping, and torture, a report confirmed in 1999 by the Guatemalan Commission on Historical Verification.[6] In November 1996, Harold J. Nicholson, a rising star in the CIA's clandestine service, was convicted of espionage and jailed for selling secrets to the Russians and exposing scores of new operations officers.[7] In May 1997, the *New York Times* reported that the CIA had admitted intentionally destroying virtually the entire record of its operation overthrowing Prime Minister Muhammad Mossadegh in Iran in 1953.[8] In January 1999, U.S. intelligence operatives were revealed to have worked undercover on teams of U.N. arms inspectors in Iraq, seriously compromising the U.N. mission.[9] Most recently, the intelligence community seems to have failed to provide adequate information on Kosovo, resulting in, among other things, the politically damaging bombing of the Chinese embassy in Belgrade.[10]

Why has there been no major reform? One reason is that Congress has traditionally seen its role not as that of a monitor or supervisor of intelligence operations, but as that of an advocate. In a telling revelation in its 1996 report, *IC 21: The Intelligence Community in the 21st Century,* the House Intelligence Committee stated, "Intelligence, unlike other federal programs, has no natural constituency; therefore, Congress plays a vital role in building public support."[11]

The Brown Commission suggested that the government may simply be unwilling to do the work necessary to regulate the intelligence community effectively, at least in terms of economics. In its report, *Preparing for the 21st Century: An Appraisal of U.S. Intelligence,* it stated, "Ultimately, the Commission concluded that developing a precise criterion for measuring the right level of intelligence resources would inevitably be too simplistic and perhaps unwise. The reality, as for many functions of government, is that intelligence capabilities are determined by whatever the nation chooses to spend on them, not by some rigorous calculation which attempts to precisely balance threats against capabilities."[12]

The Brown Commission also shrugged off any criticism of the CIA. "There have been problems," it stated, "especially at the CIA, some of which have been substantial. While these episodes are deplorable, using them to justify cutting back or eliminating intelligence capabilities is a leap the Commission is unwilling to make."[13]

The thrust of the two governmental reports was to demand, if anything, greater and more adventuresome intelligence than during the Cold War. Both reports rested on a vision of the United States using its vastly superior military and intelligence machinery to maintain its hegemony in world affairs. Hence, it would appear, the need for a greater amount of money for intelligence operations, whatever the liabilities.

Intelligence scandal and failure, as common as they have been, are not the only reasons that a major reassessment of intelligence is needed. In the post–Cold War world, the very nature of intelligence problems is shifting. This is a world in which the World Trade Center and U.S. embassies are bombed by Middle Eastern terrorists, the illegal trade of nuclear weapons and materials by bankrupt countries formerly in the Soviet Union is a continuing possibility, and countries such as Iraq and Sudan wreak international havoc and pose massive threats to human rights. It is also a world in which American trade secrets are stolen by rapacious foreign companies; international crime is on the rise; and environmental breakdowns, such as the nuclear meltdown in the Ukraine, pose grave dangers. How can these problems be dealt with? Which agencies are best equipped to deal with them? What role, if any, should the intelligence agencies play? Should these matters be handled unilaterally by the United States, or should they be dealt with in cooperation with other countries? Again, a major rethinking of our intelligence operations is prompted by the events of our recent history.

This book assembles a group of national experts, possessing firsthand experience, to provide both the facts of the intelligence establishment's failure and prescriptions for its reform in the post–Cold War world. The group comes from disparate backgrounds: the U.S. Foreign Service, the

CIA, the Office of Management and Budget, the U.S. Congress, the National Security Archive, international law, academics, and investigative journalism. Each writer views the intelligence community from his or her institutional and personal point of view. For this reason, certain incidents and facts are dealt with in more than one article, reflecting the authors' different perspectives and providing the bases for their recommendations for reform.

The experts have worked closely with the Center for International Policy, a foreign-affairs institute in Washington, D.C. The center, founded twenty-three years ago in the wake of the Vietnam war, promotes a foreign policy based on democratic values. It has played a key role in advocating the Contadora and later the Arias Peace Plans in Central America; served as advisers to the Haitian democratic government-in-exile; promoted the normalization of relations with Cuba; and sponsored intelligence seminars at the Senate Office Building, leading up to the writing of this book. (For a fuller description of the Center, see page 227.)

The book argues that today we have the strongest reasons for change, and that change can take place only through an analysis of the intelligence system from the ground up. What went wrong? How can it be fixed? What is the proper mission of intelligence? How should it be implemented?

Suggestions for change are developed by the individual authors and also emerge from the book as a whole. In Chapter 1, "After the Cold War: The Need for Intelligence," Roger Hilsman, former Assistant Secretary of State, adviser to President Kennedy, and author of *The Cuban Missile Crisis* and other works on foreign affairs, sends the reader on a tour of the post–Cold War world, reviewing region by region the strategic interests of the United States and its need for intelligence. Hilsman strongly questions whether this need is the same as it was during the Cold War.

In Chapter 2, "Espionage and Covert Action," Melvin A. Goodman, former division chief and senior analyst at the Office of Soviet Affairs of the CIA, spells out the huge price of covert operations—in financial costs, in problems concerning short- and long-term reliability, and in the creation of an international atmosphere of suspicion and violence. He also asks: Do we still have the same need for espionage and covert action that we once had?

In Chapter 3, "Too Many Spies, Too Little Intelligence," Robert E. White, former U.S. Ambassador to El Salvador and Paraguay and president of the Center for International Policy, presents a chilling personal account of CIA operations in Central America and the Caribbean.

In Chapter 4, "CIA–Foreign Service Relations," Robert V. Keeley, former U.S. Ambassador to Greece, Zimbabwe, and Mauritius, describes from personal experience the coercive tactics of the CIA toward our diplomatic service and its often destructive effects on our foreign relations.

In Chapter 5, "Covert Operations: The Blowback Problem," Jack A. Blum, an international lawyer and the former chief investigator for the Senate Foreign Relations Committee, headed by Senator Frank Church, and of the Senate investigation of the Iran-Contra scandal, explores the effects of U.S. intelligence on this country. The chapter deals with the erosion of respect for democracy, increases in violence and drugs, and the adoption of intelligence methods in domestic politics.

In Chapter 6, "The End of Secrecy: U.S. National Security and the New Openness Movement," Kate Doyle, an analyst at the National Security Archive, describes how the intelligence agencies continue to hide behind screens of "national security" and "protection of sources and methods," which deny citizens access to records and deny policymakers and the public access to our history.

In Chapter 7, "Mission Myopia: Narcotics as Fallout From the CIA's Covert Wars," Alfred W. McCoy, author of *The Politics of Heroin: CIA Complicity in the Global Drug Trade,* spells out how U.S. intelligence has promoted drug dealing around the world, most recently in the United States itself.

In Chapter 8, "TECHINT: The NSA, the NRO, and NIMA," Robert Dreyfuss, a journalistic expert on electronic intelligence, describes the huge costs and limitations of electronic intelligence as well as its clear uses. He also depicts the compromising relationships among corporate contractors, the intelligence community, and congressional committees.

In Chapter 9, "Improving the Output of Intelligence: Priorities, Managerial Changes, and Funding," Richard A. Stubbing, currently at Duke University and formerly in charge of the intelligence budget at the U.S. Office of Management and Budget, presents a series of recommendations for improving the management of intelligence and its economic accountability.

In Chapter 10, "Who's Watching the Store? Executive-Branch and Congressional Surveillance," Pat M. Holt, former chief of staff of the Senate Foreign Relations Committee and the author of *Secret Intelligence and Public Policy,* explores the problems and possibilities of altering the mission, accountability, and supervision of intelligence.

In "Conclusions," I draw together the recommendations for reform that end each of our authors' chapters. Perhaps the most extraordinary fact that has emerged in the editing of this book is the authors' agreement on the problems facing U.S. intelligence and what needs to be done to improve it. Each author has examined different aspects of the operations of the intelligence community, and yet when the recommendations emerging from this book are seen together, they form a remarkably coherent and unified plan for reform. Reform of U.S. intelligence is not an unsolvable mystery, but emerges as a clearly articulated goal resulting from dispassionate analysis, untainted by institutional or bureaucratic partisanship and informed by concern for the national interest.

Intelligence, because of its secret nature, demands that the government itself institute reforms. But with the continued failure of the government to act, the country is left no alternative but to exert public pressure and put the government on notice that it is expected to do the job it has for so long resisted. It is the hope of the contributors to this volume, and that of its editor, that the book will spark public interest in this vital area of national policy and help provide the impetus for reforms long overdue, to give the United States an intelligence system that, while providing for national security, is commensurate with the ideals of our democratic system.[14]

NOTES

Acknowledgments: The Center for International Policy wishes to acknowledge the help of David F. Rudgers, formerly a senior analyst of the CIA, who reviewed the entire manuscript and provided useful suggestions and comments (see Bibliography); the valuable assistance of the center's James Morrell, who also reviewed the manuscript; the careful fact-checking of Jacqueline McCafferty; and the inspired editing of Hal Goodman for Temple University Press.

1. *New York Times,* October 21, 1998.
2. See, for example, "What 'New' Role for the C.I.A.?" by Director of Central Intelligence George J. Tenet, *New York Times,* October 27, 1998.
3. *New York Times,* December 4, 1998.
4. *New York Times,* September 28, 1994.
5. Commission on the Roles and Capabilities of the United States Intelligence Community, *Preparing for the 21st Century: An Appraisal of U.S. Intelligence* (Washington, DC: Government Printing Office, March 1, 1996); House Permanent Select Committee on Intelligence, *IC 21: The Intelligence Community in the 21st Century* (Washington, DC: Government Printing Office, April 9, 1996).
6. *New York Times,* June 29, 1996; February 26, 1999.
7. *New York Times,* November 19, 1996. See also the case of David Sheldon Boone, an NSA analyst arrested for spying for the Soviets, reported in the *New York Times,* October 14, 1998.
8. *New York Times,* May 31, 1997.
9. *New York Times,* January 7, 1999.
10. *New York Times,* May 9, 1999.
11. *IC 21,* p. 28.
12. *Preparing for the 21st Century,* p. 134.
13. *Preparing for the 21st Century,* p. 10.
14. See my "Bringing Light to Covert Operations" in the *Baltimore Sun,* August 21, 1995, and a subsequent exchange of letters with John Deutch, director of the CIA, on September 28 and October 7.

ROGER HILSMAN

1 After the Cold War
The Need for Intelligence

THE PARAMOUNT ROLE of intelligence is to supply the President and Congress with information about possible strategic threats to the United States.[1] During the Cold War, this information was supplied by U.S. embassies and consulates, by the foreign correspondents of the American news media, by the newspapers and broadcasts of other countries, and by espionage. The United States attempted to influence the actions of other countries by the representations of its Ambassadors and consular officers, by the public pronouncements of the President and other American officials, by acts and resolutions of Congress, and by covert political actions—defined as efforts to shape events in other countries by secret or at least "plausibly deniable" means. The question is: Should essential gathering of information and influencing of actions rely as much on espionage and covert political action in the future as they did during the Cold War?

During the Cold War, more than 60 percent of the U.S. intelligence effort focused on the Soviet Union. The stakes were extremely high. The Soviet Union was an empire ideologically committed to replacing our system of government. It possessed a nuclear strike force more than adequate to destroy the major population centers of the United States, and a conventional military force several times larger than ours. With a government schooled by a centuries-old tradition of secrecy, stretching back to Peter the Great, it presented a formidable challenge to U.S. intelligence.

As we will see, U.S. intelligence's record of success was mixed. Aided by satellites, it often gave U.S. policymakers clear information on specific targets and distributions of enemy forces. At the same time, it frequently exaggerated enemy force levels, and thereby distorted U.S. policy. Despite an enormous effort to penetrate Soviet intelligence and policymaking, its successes were limited. In the end, it failed to predict one of the most momentous political events of the century: the breakup of the Soviet Union into Russia and fourteen independent nations.

U.S. intelligence had been created with the precise purpose of dealing with this formidable adversary. To do so, it employed almost every conceivable means, including election-rigging, political subversion, and the use of paramilitary force. Its operations were carried out not only in the

Soviet Union and Eastern Europe but in the rest of the world as well. In the neutral world, U.S. intelligence competed on a daily basis with Soviet and Eastern European intelligence agents. In areas within the U.S. "sphere of influence," such as Western Europe and Central and South America, U.S. intelligence stopped at nothing to insure that no piece of territory would be lost or compromised.

At home, the U.S. intelligence agencies shrouded themselves in protective secrecy, carrying out operations without the usual democratic protections of openness. At the same time, as exposed by the Church Committee, the agencies repeatedly violated the civil rights of U.S. citizens when they deemed this necessary.

Viewed retrospectively, U.S. intelligence made only a minimal contribution to the demise of the Soviet Union in 1989. It seems clear that the Soviet Union broke up not because of the operation of U.S. intelligence, but because its people were fed up with repression, because the Soviet economy simply could not compete with the economy of the United States, and because of the intense nationalism of components of the Soviet Union, which sought desperately to free themselves from Moscow's rule. The huge investment that the United States made in intelligence did virtually nothing to end communism. It was useful in monitoring Soviet force levels and nuclear capability, but it was ineffective in bringing about political change. By going beyond its primary mission of supplying the President and Congress with information about strategic threats to the United States, it attempted futilely, through espionage and covert action, to create a political result that, when it came, was the product of entirely different forces.

STRATEGIC THREATS AND THE NEED FOR INTELLIGENCE TODAY

The world today, including the former Soviet Union, presents the United States with a significantly diminished threat and vastly different intelligence problems than it did during the Cold War. It therefore seems clear that our intelligence system should be modified. The following pages undertake a brief world tour, assessing threats to the United States and the intelligence needs they create. The purpose is to establish the basis for a redesign of the intelligence system for the contemporary world, not for the past world of the Cold War.

The Former Soviet Union

After the breakup of the Soviet Union, the successor states agreed to turn over all of the nuclear weapons in their territories to Russia. In 1992,

George Bush and Boris Yeltsin agreed to reduce the long-range missile warheads in their stockpiles to between 3,000 and 3,500 each. A quick calculation indicates that it would take about 200 warheads to destroy the seventy metropolitan areas that contain more than 50 percent of the American population, and about 300 warheads to destroy the much larger number of urban areas containing more than 50 percent of the Russian population. In other words, the importance of the reduction was only symbolic. A sudden turn of events, such as we have seen in Kosovo, could easily put the United States and the new Russia into tension that could bring them back to the Cold War—or to the kind of nuclear-missile crisis that the world saw in Cuba in 1962.

One element, however, is crucially different. The Soviet Union had a "nationalities problem" that required the careful attention of the central government. But dealing with rival nationalities within a unitary state is much, much easier than dealing with rival nationalities that are states themselves. When these nationalities were part of the Soviet Union, fear of Germany and later of the United States helped keep them quiet. That's no longer a factor. More important, it's vastly easier for a rival nationality that is itself a state to raise troops and equip them with arms. The conclusion seems inescapable that the new Russia will be much more concerned with its new neighbors than with the United States and the rest of the world. Today's Russia obviously has no ambitions that threaten Western Europe or the United States, but since it has both the missiles to reach Europe *and* the United States and the warheads to make them nuclear, there continues to be a need for intelligence on developments inside Russia that might give it such ambitions. Here again, what is required is continued intelligence about strategic forces, not ambitious schemes for overthrowing governments through espionage or covert action.

Eastern Europe

The United States will need to watch carefully for any signs that the mutual fears and rivalries in Eastern Europe may turn violent, escalate, and eventually drag the West into the conflict. The expansion of NATO to include the Czech Republic, Poland, and Hungary goes a long way toward neutralizing the possibility. The expansion did cause some alarm in Russia, but a major speech addressing the question by Under Secretary of State Strobe Talbott in February 1998 helped quiet fears. Recent tensions over Kosovo, however, suggest that such fears can easily reappear, with possibly dangerous consequences.

Asia

In Asia, there continues to be tension between Japan and Russia over the Kuril Islands, annexed by the Soviet Union after World War II. But it is not a tension that is likely to lead to war or to impinge on vital American interests.

As for China, its surge of economic growth and new prosperity reduces the incentive for adventurism of any kind. China is in social, political, and economic turmoil, but the turmoil is internal.

Tension between Communist China and the Republic of China on Taiwan seems to be contained, but it does continue and could easily take a dangerous turn. In addition, China has for some time had a stockpile of nuclear warheads and a growing ICBM arsenal, including missiles aimed at the United States. Obviously, America needs to maintain a careful watch on developments in China to make sure that it continues to be preoccupied with internal affairs.

North Korea has pursued an extensive program to develop nuclear warheads, but it agreed in 1994 to give up its nuclear-weapons program in exchange for billions of dollars in Western aid. Shortly thereafter, Kim Il Sung, the longtime dictator, died. He was succeeded by his son, Kim Jong Il. Doubts about the son's intentions to honor the agreement rose when North Korea sold missiles to Pakistan and infiltrated nine commandos into South Korea by submarine. Then, in the summer of 1998, United States intelligence detected a huge, secret underground complex in North Korea that it believes is the centerpiece of an effort to restart the nuclear-weapons program. North Korea has also developed missiles capable of carrying conventional, chemical, or nuclear warheads with the still untested capacity to hit Hawaii and Alaska's western fringe. Clearly, close surveillance of North Korea is vital.

In Southeast Asia, there are tensions between Vietnam and China and between Cambodia and Vietnam. Neither situation is likely to result in a war that might spread to affect U.S. interests, but they bear watching.

In South Asia, tension between India and Pakistan is high, aggravated by the successful nuclear tests recently conducted by both countries and by confrontation in Kashmir. There is evidence that both have missile-launching capacity, specifically India's Agni and Pakistan's Ghauri missiles. It is also worth noting that even though the nationalist Indian political party that recently came to office was committed to making India a nuclear power, and even though the United States was aware of unusual activities at the Indian test site, the CIA concluded that no tests would be conducted. It was a major intelligence failure.

The governments and people of both India and Pakistan, by promising to negotiate an end to tests, have shown that they are sensitive to the consequences of war and the possibility that it would become nuclear. However, if these promises come to nothing, if war does come, and if that war becomes nuclear, it is worth noting that such a war would have little potential for escalating into a conflict that would threaten world peace or draw in either Russia or the United States. Even so, the consequences of an India-Pakistan war would be severe, and the situation requires careful attention.

Although fighting continues between the different factions in Afghanistan, the struggle is not likely to escalate. On the other hand, tension between Iran and Afghanistan rose following Afghanistan's killing of a number of Iranian diplomats. The situation in Afghanistan obviously bears watching.

The Middle East

Tensions with the potential for escalating into war abound. First and foremost is the tension between Israel and the Palestinians, followed by tensions between Israel and Syria, and Israel and Egypt. Iran and Iraq fought a bloody war for several years that ended in a stalemate. Iraq invaded Kuwait and the Gulf War followed. Tensions continue between Egypt and Saudi Arabia, Iraq and Saudi Arabia, and Iraq and Turkey.

The Case of Iraq

Iraq presents a special problem. In 1981, Israeli intelligence became convinced that Iraq was secretly attempting to develop an atomic bomb, and Israel bombed the installation at Twaitha where the work was going on. Although this bombing probably set back the Iraqi attempt to build nuclear weapons, Iraq immediately started a new, more tightly guarded, and much more extensive program.

As part of the agreement that ended the Gulf War, Iraq was forced to allow teams of U.N. inspectors to investigate suspected nuclear-weapons plants and facilities. A cat-and-mouse game ensued, with Iraq attempting to hide materials and equipment related to its nuclear program and the U.N. teams trying to sniff them out.

Gradually, the U.N. teams built up a picture of an Iraqi nuclear program that was both larger and more sophisticated than anyone in the West had suspected. The program employed more than 10,000 scientists and technical workers, and had cost at least $10 billion. American and allied bombing has destroyed a lot of the brick-and-mortar installations, but many key materials and much equipment have survived.

More important, Iraq's formidable corps of scientists, technicians, and weapons experts has been unharmed. Iraq, in fact, has the largest technical base in the Middle East. Although Israel's is qualitatively better, Iraq's is larger.

The evidence gathered by the U.N. teams suggested not only that Iraq could have tested a crude fission bomb within a year, but that it was trying to develop the capacity for a hydrogen bomb. Iraq had a stock of deuterium oxide, so-called heavy water, that it had purchased some years earlier. Hans Blix, head of the International Atomic Energy Agency, said that his inspectors had found documentary evidence that Iraq was planning to produce lithium-6, whose only known use is in H-bombs.[2] When deuterium oxide and lithium-6 are combined, the product is lithium-6 deuteride, the main component of an H-bomb. The bottom line is that if Iraq finally succeeded in using these materials to make an atomic bomb that was not too crude or heavy, it might be able to transform it into a hydrogen warhead small enough to be delivered by an airplane or even by missile. U.N. officials said that Iraq had tested a missile that would be able to carry a warhead of the necessary size, and that it could reach Turkey, Saudi Arabia, or Israel. Similar warheads carried by American missiles have the explosive power of 300 kilotons of TNT, as compared with the fourteen kilotons of the Hiroshima bomb. Experts estimated that had it not been for the Gulf War, Iraq could have tested a reasonably sophisticated atomic device by 1993 or 1994, a fission bomb some time later, and an H-bomb several years after that. However, it would still need to develop delivery capability, which in some ways is a more difficult problem than building a bomb.

What is clear is that American and allied intelligence underestimated the size of the Iraqi nuclear program and overestimated the damage that the bombing during the Gulf War had done to it. On January 23, 1991, for example, President Bush said, "Our pinpoint attacks have put Saddam out of the nuclear bomb-building business for a long time."[3] On several different occasions, General H. Norman Schwarzkopf, the American commander of the allied forces, said that the bombing attacks "had destroyed all their nuclear-reactor facilities" and "neutralized their nuclear manufacturing capability."[4]

After the war, high-ranking Pentagon officials admitted to the press that the intelligence failure concerning Iraq's nuclear program had been extensive and that the experience raised serious concerns about how much the United States could learn about the nuclear program of any potentially hostile country, including Libya and North Korea as well as Iraq.

The Case of Iran

In July 1998, Iran tested an MRBM (Medium Range Ballistic Missile) with a range of 800 miles. This made it capable of reaching all of Israel, Iraq, Afghanistan, and Pakistan, most of Saudi Arabia and Turkey, and part of Russia. The missile—one of a number that Iran had bought from North Korea and named Shahab 3—blew up in the latter stages of its flight, but it may well have been detonated deliberately. Iran also bought the technology and equipment to make a nuclear warhead for such missiles from China, but Western intelligence believes that it will be several years before it can be tested.

In December 1997, the Clinton administration certified that China was no longer helping Iran, Pakistan, or any other country build a nuclear bomb and that China was ending its nuclear-power projects with Iran, which it said were not related to weapons. In March 1998, the United States offered Russia an opportunity to expand its lucrative business in launching American satellites if Moscow would clamp down on its sales of missile technology to Iran. Nevertheless, Iran obviously bears watching.

Israel

Israel has a stockpile of nuclear warheads and the missiles to deliver them anywhere in the immediate vicinity. The most serious danger is a war between Israel and the Arab world that goes so badly for Israel that it begins to believe that it will be engulfed. In such circumstances, Israel might well exercise the "Samson option" and bring the whole of the Middle East tumbling down, as Samson did the temple.

Still, the handshake between the Arab and Israeli leaders on the White House lawn and the subsequent negotiations between Israel and the Palestinians, although disappointing, have helped to quiet such fears. And both sides clearly understand how horrendous the consequences would be if events spiraled out of control.

Algeria

Concern about Algeria's nuclear potential is great. It was well known in the West that the Chinese had supplied Algeria with a nuclear reactor. Many U.S. and other officials suspected that it was intended not for peaceful purposes but for research on nuclear weapons and eventually their production.

To sum up, the conclusion is obvious that the intelligence need in the Middle East is great if the President and Congress are to be adequately informed of nuclear threats, though these are minor in comparison with the strategic threats posed by the Soviet Union during the Cold War.

THE THREAT OF TERRORISM

Conventional thinking about foreign relations and the proper role of intelligence must take into account the substantial increase in terrorism, which presents new demands on our intelligence system.

Terrorist States

In July 1998, a panel headed by Donald H. Rumsfeld, former Secretary of Defense, said that North Korea, Iran, and Iraq could develop nuclear missiles capable of striking the United States with little or no warning. In response, the CIA said that it stood by its estimate that no nation besides Russia and China could field long-range missiles before the year 2010.

Suppose, however, that North Korea, Iran, or Iraq, without acquiring long-range missiles, somehow obtained half a dozen nuclear warheads. In a war with one of its neighbors using short-range missiles, aircraft, or secret agents with "suitcase" bombs, it could destroy the larger cities of its enemy, which might well make the difference between defeat and victory. The long and bloody war between Saddam Hussein's Iraq and the Ayatollah Khomeini's Iran ended in stalemate, but if one side had had a dozen nuclear weapons it could have won.

To attack the United States, a terrorist state that lacked long-range bombers or missiles would have to use secret agents with suitcase bombs. Terrorist organizations used conventional bombs on a disco in Germany frequented by American soldiers, against the World Trade Center in New York, and against the American embassies in Kenya and Tanzania. Terrorist organizations are unlikely to acquire nuclear weapons, but suppose a terrorist state used a nuclear weapon against such a target, either out of hate or in an attempt to provoke a war between the United States and another nuclear power—Russia, for example. As many as several million people might be killed. But it seems doubtful that the attack would succeed in provoking a war between the United States and Russia, if that was the motive. In fact, concealing the evidence that would lead back to the true culprit would be almost impossible, and when the United States and Russia found out who the culprit was, they would almost certainly take joint action to end forever the threat from that state.

Terrorist Organizations

As mentioned above, the possibility that a terrorist organization, as opposed to a terrorist state, could acquire a nuclear weapon seems remote, but it is not inconceivable. It might be given a weapon by a pariah state sympathetic to its views, it might bribe its way into possessing one, or it might conceivably steal one. It is difficult to see what advantages over ordinary explosives a suitcase nuclear weapon would offer to the leaders of a terrorist organization, but it's possible that fanatics would see an advantage in the drama of using a nuclear rather than a conventional weapon.

What seems extremely unlikely is that a terrorist organization could get its hands on more than one or two nuclear devices. This means that casualties would be in the range of thousands or tens of thousands, rather than millions. As McGeorge Bundy pointed out, even a very successful terrorist attack with nuclear weapons would be no worse than the Chernobyl accident.[5]

In the short run, the only answer to the threat of nuclear terrorism by either an organization or a pariah state is very effective intelligence and police work and very tight security. Since even the best of such measures cannot guarantee total protection, it may be cold comfort that this is one area in which all the great powers have an extremely large stake in cooperating fully, but it is still true.

In the meantime, it might be helpful for the great powers to convene a conference on this subject with the aim of making a public announcement that they will invade and destroy any state that uses or permits a terrorist organization based on its territory to use nuclear weapons.

Chemical and Biological Weapons

Many people both inside and outside the American government believe that the most immediate threat from terrorists is that they will acquire not nuclear weapons but chemical and biological ones. In May 1994, the judge in the World Trade Center bombing trial announced that the defendants had put sodium cyanide in their explosive package. If so, it must have burned up in the explosion; otherwise it could have created a cyanide gas cloud that might have poisoned thousands of people. In March 1995, two members of the Minnesota Patriots Council, a militia-type organization, were convicted of planning to use ricin to assassinate federal agents. A tiny bit of ricin can kill in a matter of minutes if ingested, inhaled, or absorbed through the skin. Also in March 1995, the Japanese doomsday cult Aum Shinrikyo released sarin nerve gas in a Tokyo subway at the height of the rush hour, killing twelve people and hospitalizing more than 5,000. In the summer of 1995, Iraqi defectors revealed that Iraq had a massive chemical- and biological-weapons program that included anthrax, botulinum, sarin, and VX, the most lethal form of nerve gas. In December 1995, a man with alleged ties to survivalist groups was charged with attempting to smuggle 130 grams of ricin into the United States.

The only way to prevent terrorists from getting weapons of mass destruction, whether nuclear, chemical, or biological, is tight security and effective intelligence work. The result of even a very successful attack would be more like the nuclear meltdown at Chernobyl than Armageddon. Even so, a then-secret meeting in the White House in March 1998 concluded that the United States is currently unable to handle a germ-warfare attack, and on May 22 President Clinton announced a program against what he called "biological, computer, and other 21st-century threats."[6]

The need for intelligence on both foreign countries and terrorist groups, to sum up, is substantial. What about the means for acquiring it?

Do We Still Need Cold War–Style Intelligence?

Cryptography

When Henry L. Stimson took over as Secretary of State in 1929, he discovered that the State Department had a "Black Chamber" that had succeeded in breaking Japan's codes. The product had been very useful during the negotiations on the London Naval Treaty, but Stimson decided to abolish the Black Chamber anyway. In explanation, he later remarked that "Gentlemen do not read each other's mail."[7]

In World War II, when gentlemen were not so easy to find, Stimson, who was then Secretary of War, had no objections to code-breaking. And indeed, there should be none today. When it is successful, the product is often very useful, and there are no risks involved. With the development of computers, breaking codes is much more difficult than it once was, but clearly the National Security Agency should continue to try. It should also continue its work on protecting our own codes.

Aerial and Satellite Intelligence

During the Cold War, American planes loaded with electronic devices probed the perimeters of the Soviet Union to learn everything possible about Soviet radar. Various satellite devices orbited over the Soviet Union to monitor other forms of electronic data, and U-2 planes flew over the country taking marvelously detailed pictures of military and industrial installations.

The U-2 flew above the reach of Soviet antiaircraft defenses, and it was extraordinarily successful until May 1, 1960, when Gary Powers was shot down by a greatly improved Soviet surface-to-air antiaircraft missile. Within a couple of years, a successor to the U-2, the SR-71 Blackbird, began flying. It operated from 1964 until 1990, when it was retired, presumably because it was felt that unmanned satellites could do the job just as well without risk. Subsequently the SR-71 was returned to active duty for some unspecified, limited task.

Both the U-2 and the SR-71 violated the airspace of other countries. The international consensus today, however, is that there has to be some limit on how far into space national control should extend, and that such restrictions should not apply to satellites, whether they are taking pictures or eavesdropping on electronic communications.

Clearly, the United States should continue satellite reconnaissance in all its forms. It is an extremely valuable source of information, allowing the United States to monitor nuclear-launch capability with minimal intrusiveness, and it has been legitimized in international law. Given the possibilities we have reviewed that potentially hostile countries may develop nuclear weapons, satellite reconnaissance seems fully justified and necessary.

Espionage

The most important thing to be said about espionage, the stuff of spy-thriller fiction, is that even though the "take" from espionage can be extremely valuable, only a very tiny percentage of the information gathered by the United States comes from espionage, and only very rarely is it decisive. The reasons are obvious. First, placing and maintaining agents in spots where they can gather decisive information is extraordinarily delicate and difficult. Second, communicating with such agents after they are in place is intricate and time-consuming, and the cost in lives and treasure is very high.

So are other, intangible costs. Almost every American embassy in the world has its collection of CIA intelligence officers. In some countries, their role is merely that of liaison officers with the particular country's own intelligence organizations. But in other countries—even friendly ones—it is much more sinister. Over time, the potential effect of maintaining a CIA component in an American embassy can be corrosive.

One example will suffice. In Singapore during the Eisenhower administration, the main responsibility of the CIA station in the embassy was to serve as a liaison and exchange intelligence information with the CIA's Singaporean counterpart, the Special Branch, which was headed by officers on loan from Great Britain. But the local CIA officers saw an opportunity to subvert a member of the Special Branch, and the Washington headquarters could not resist the temptation to have "an independent check on whether information was being fully shared." When the Kennedy administration came to office it found that a CIA man from the Tokyo station was in jail in Singapore, having been arrested a few weeks earlier in the act of attempted subversion. With great difficulty the Kennedy administration persuaded Singapore's Prime Minister, Lee Kuan Yew, to let the CIA man go without any publicity. But years later, when tension arose between Singapore and the United States during another administration, Lee Kuan Yew could not resist the temptation to make the affair public in a bitter denunciation.

To sum up, the cost of espionage has been high—not so much in terms of lives lost (although a number of American and foreign lives *have* been lost) as in terms of treasure and, especially, America's reputation.

The gain from espionage, however, has not been great. The most convincing illustration is probably the period that witnessed the breakup of the Soviet Union. The result of Aldrich Ames's treason in exposing CIA contacts in the Soviet Union was that the United States had no agents inside that country during the period of its breakup. Even worse, the Soviets had succeeded in planting double agents in various places; these agents fed the CIA misleading information, which the Agency duly passed on to the top U.S. decision-makers. Although no one at the top levels of the U.S. government foresaw the rapid and tumultuous events, neither did Mikhail Gorbachev and the leaders of the Soviet Union, who can hardly be said to have lacked information.

The fact is that the American embassy, newspaper reports, and the other open means of acquiring information kept American leaders reasonably well informed. The obvious conclusion is that espionage would have made very little difference in the way the United States reacted to events during the last few years of the Soviet Union. Even if it had made a difference, it's extremely doubtful that the benefit would have been worth the cost and effort of recruiting, maintaining, and communicating with the agents involved. With the end of the Cold War and the breakup of the Soviet Union, espionage is something that the United States can do without.

Covert Action

The most prominent advocate of covert political action was Allen W. Dulles, who had served in the Office of Strategic Services during World War II and was Director of Central Intelligence from 1953 to 1961. Citing the Truman and Eisenhower doctrines, which had laid down the policy that the United States would come to the aid of countries threatened by communism whose governments requested help, Dulles enunciated a doctrine of his own. He argued that covert political action should be used to foil communist attempts to take over a country *with or without a request for help*. Thus, Dulles wrote: "In Iran a Mossadegh and in Guatemala an Arbenz came to power through the usual processes of government, and not by any communist coup as in Czechoslovakia. Neither man at the time disclosed the intention of creating a communist state. When this purpose became clear, support from outside [by the CIA] was given to loyal anti-communist elements in the respective countries.... In each case the danger was successfully met. There again no invitation was extended by the *government* in power for outside help."[8]

In neither case, of course, was Dulles correct in his justification: that Mossadegh and Arbenz were committed to creating communist states. What seems clear is that in each case Dulles radically extended both his definition of communism and the perimeters of the Cold War to justify actions that retrospectively can be seen to have been of dubious value to the United States. (See Chapters 2 and 3.) The option of covert action, tempting as it may be, is fraught with danger.

In abstract and theoretical terms, Dulles's position is not unreasonable. If a great power—communist, Nazi, militarist, imperialist, or whatever—is openly antagonistic to the rest of the world, and if this hostile power uses the techniques of subversion to bring down foreign governments, then the countries that are the targets of this hostility have both a right and a duty to defend themselves by whatever means are necessary, including covert political action. Or at least, they have a right and a duty to defend themselves with such methods *where those methods are effective and appropriate, and when there is no effective and appropriate alternative.*

The trouble has been that these two qualifications have not always been observed. In the past, the United States has too often used secret intelligence methods when they were neither effective nor appropriate, and when there were effective and appropriate alternatives. Covert political action became an obsession—the answer to every kind of problem—and U.S. agents became as ubiquitous and busy as those of the communists. Where statecraft and an imaginative and open use of our resources might have achieved desirable results, the United States attempted, often with disastrous consequences, to achieve short-term gains through stealth and subversion. Unfortunately, these means carry their own message, and they often corrupt the very situations they are designed to correct.

In the years following the establishment of the CIA, and especially after General Walter Bedell Smith made the CIA a more effective organization during his tenure as DCI from 1950 to '53, the United States first came up against some of the perplexities of the postwar world and the communist threat. It met the direct communist challenge by building up its military strength and by alliances. But again and again, the United States had to face problems that did not yield to power alone: ineffective governments; graft; politically apathetic populations; indifferent leaders; and communist subversion, terrorism, and guerrilla warfare.

There were also instances in which the world would have regarded the use of naked power as petty bullying. Everyone agreed at the time of the Bay of Pigs fiasco, for example, that it would be politically disastrous if the United States were to attack tiny Cuba directly and overtly. To many of the policymakers who faced such problems, covert political action seemed a way out. The United States would fight communist fire with fire.

Since the action was "covert," it promised to get around the obstacles, the moral problem of "intervention," and the political problem of appearing to be a bully. Covert action also had an aura of omnipotence, as if, like James Bond, the fictional Agent 007, it could accomplish miracles.

But covert action was really just a gimmick. In very special circumstances it might be a useful supplement, but nothing more. It is one thing to give a covert boost to, for instance, the Philippines' Magsaysay, a natural leader with wide popular support, in a bid for power in the midst of a national crisis created by the guerrilla terrorism of the communist Hukbalahaps. But it is something quite different to try to create a Magsaysay by covert efforts, as the United States did in the 1950s with General Phoumi Nosavan in Laos.

President Truman, who had finally acquiesced in the establishment of the CIA, later expressed his misgivings that the Agency had been allowed to engage in covert political action. "For some time," he wrote in 1963, "I have been disturbed by the way CIA has been diverted from its original assignment. It has become an operational and at times a policy-making arm of the government."

Truman went on to say that he never had any thought when he set up the CIA that it would engage in peacetime cloak-and-dagger operations, but that he had intended it to be confined to intelligence work. He wrote, "Some of the complications and embarrassment that I think we have experienced are in part attributable to the fact that this quiet intelligence arm of the President has been so removed from its intended role that it is being interpreted as a symbol of sinister and mysterious foreign intrigue—and a subject for Cold War enemy propaganda."

Truman's conclusion was that he would like to see the CIA restored to its original assignment as the intelligence arm of the President. "We have grown up as a nation," he wrote, "respected for our free institutions and for our ability to maintain a free and open society. There is something about the way the CIA has been functioning that is casting a shadow over our historic position and I feel that we need to correct it."[9]

Covert action, it seems clear, *has* been overused as an instrument of foreign policy, and the reputation of the United States has suffered. "Covert political action" is usually defined not as completely secret but as "plausibly deniable." But whereas one action may be plausibly deniable, several hundred are not. And although one action, considered in isolation, could seem worth the cost of a slight tarnishing of the American image, the cumulative effect of several hundred blots has been to blacken it entirely. As early as the end of the Eisenhower administration, America's reputation was such that the CIA got credit for almost everything unpleasant that happened in many parts of the world, whether or not it was actually responsible. Relying too heavily on the techniques of secret intelligence so corroded one of America's major political assets—the belief in American integrity—as to nullify any possible gain.

Covert political action is not only something the United States can do without after the Cold War, it's something the United States could have done without during the Cold War.

If the United States gets out of the business of espionage and covert political action, it could eliminate all the CIA contingents it now maintains in its embassies all over the world, except for those engaged in liaison with their intelligence counterparts from other countries. The dollar savings would be substantial and the gain in political terms would be greater still.

With the end of the Cold War and the breakup of the Soviet Union, the American intelligence community needs to be reorganized. There is more to be gained than lost by getting out of the business of espionage and covert political action, and by closing out all the CIA stations in American embassies abroad, except for those engaged in liaison with allies. Satellite observation continues to be extremely valuable and entails no political cost. The same is true of monitoring foreign broadcasts and of the codebreaking and monitoring performed by the National Security Agency.

The details can be debated. But it seems very obvious indeed that the CIA and the entire intelligence community should be subjected to a careful, thorough examination, and that some fundamental decisions be made about the kind of intelligence organizations the country needs in an age no longer dominated by a Cold War with the Soviet Union.

NOTES

1. See the following of my books: *Strategic Intelligence and National Decisions* (1956; reprint, Westport, CT: Greenwood Press, 1981); *George Bush vs. Saddam Hussein, Military Success! Political Failure?* (Novato, CA: Lyford Books, Presidio Press, 1992); *The Cuban Missile Crisis* (Westport, CT: Greenwood Publishing Group, 1996); *From Nuclear Military Strategy to a World Without War: A History and Proposal* (Westport, CT: Praeger, 1999). Also various articles on intelligence, especially "Does the CIA Still Have a Role?" *Foreign Affairs*, September/October 1995.

2. *New York Times*, February 6, 1992.

3. *New York Times*, January 24, 1991, p. A14.

4. In two television interviews on January 20, 1991, as reported in the *New York Times*, January 21, 1991.

5. McGeorge Bundy, *Danger and Survival: Choices About the Bomb in the First Fifty Years* (New York: Random House, 1988), p. viii.

6. *New York Times*, May 23, 1998, p. A18.

7. Henry L. Stimson and McGeorge Bundy, *On Active Service in Peace and War* (1948; reprint, New York: Harper & Brothers, 1971), p. 188.

8. Allen W. Dulles, *The Craft of Intelligence* (New York: Harper & Row, 1963), p. 224.

9. Harry S. Truman, in an article syndicated by the North American Newspaper alliance, as it appeared in the *Washington Post*, December 22, 1963.

MELVIN A. GOODMAN

2 Espionage and Covert Action

> *In the history of nations the influence of spying has been generally exaggerated. It is true that the secret services of states have played exciting underhand roles throughout modern history. But their clandestine activities were seldom formative or decisive: what most of the dramatic achievements of secret agents amounts to is the gathering of precious fragments of information that may or may not confirm but that does not formulate already existing diplomatic and strategic policies.*
>
> John Lukacs, *The End of the 20th Century and the End of the Modern Age*

COVERT ACTION, in the U.S. intelligence lexicon, refers to a secret operation to influence governments, events, organizations, or persons in support of a foreign policy in a manner that is not attributable to the United States. These actions may include political, economic, propagandistic, or paramilitary activities. The term "covert action" is a peculiarly American invention; it does not appear in the lexicons of other intelligence services. Nor does the term appear in the National Security Act of 1947, which created the Central Intelligence Agency, the Department of Defense, and the National Security Council.

All postwar Presidents have used the CIA for covert action as well as for intelligence collection and analysis; within months of its creation, the CIA was engaged in sensitive covert actions in Western Europe, influencing elections in Italy and France. At the direction of the National Security Council and the urging of Secretary of State George C. Marshall, the CIA provided funds to the Christian Democratic Party and to anticommunist trade unions in Italy in order to prevent a communist victory in the 1948 presidential elections. The CIA helped to fund similar campaigns in France and Japan; these "black bag" payments usually were concealed from the U.S. Ambassador. The early successes led to even wider use (and misuse) of covert action, including assassination plots against Fidel Castro and the disastrous Bay of Pigs invasion in 1961, which damaged the presidency of John F. Kennedy, and the illegal use of federal funds in the Iran-Contra scandal of the 1980s, which did the same to Ronald

Reagan. Dwight Eisenhower, Richard Nixon, and Jimmy Carter considered covert action a major instrument of policy. Other Presidents, such as Lyndon Johnson and Bill Clinton, resorted to covert action but remained highly suspicious of the CIA.

For nearly 45 years, clandestine operations played a major role in the collection of intelligence in the Soviet Union and Eastern Europe in support of U.S. policy against Moscow. Covert action captured the imagination of the CIA's political masters in the White House and Congress and, according to Robert M. Gates, former director of the Agency, these activities were the CIA's "heart and soul."[1] All covert actions were conducted on orders from the White House, according to recent congressional investigations, as well as those headed by Senator Frank Church and Representative Otis Pike in the mid-1970s.[2]

Today we must consider whether clandestine operations should play the same prominent role in U.S. foreign policy. The end of the Cold War and the collapse of the Soviet Union clearly demand a reexamination. Covert action and even clandestine collection can seriously damage U.S. bilateral relations and are often at odds with U.S. values. Since 1995, CIA officers have been embroiled in public accusations of spying by France, Germany, India, Italy, and Japan, and Agency officials concede that the "tradecraft" of their agents in recent years has been less than professional. The failure to detect Indian nuclear testing in 1998 was linked to the inept performance of the CIA station in New Delhi, and the U.S. bombing of a pharmaceutical factory in Sudan in the same year raised serious questions about the methodology of clandestine collection of intelligence used to justify military force.[3]

This discussion of clandestine operations will try to answer the following questions:

- What are the long-term ramifications of covert actions?
- What is the potential for "blowback," or negative fallout, from these operations?
- Have paramilitary operations contributed to U.S. security?
- What can be done about the isolation of the CIA's Directorate of Operations (DO) from the national-security culture in Washington?
- What are the national-security reasons for espionage and covert action in the post–Cold War era? Are policymakers relying too heavily on covert action?
- Do clandestine operations affect the CIA's ability to provide objective intelligence analysis?

A Legacy of the Cold War

> The salient lesson that I learned out of Iran-Contra was that other parts of the intelligence community can cause controversy, but it seems like the clandestine service is the only part that can cause real trouble.
>
> Robert M. Gates, 1991[4]

CIA clandestine activities—both covert action and, to a lesser extent, clandestine collection—grew out of an exaggerated notion of the threat to our security during the Cold War. The sudden and unexpected collapse of the Warsaw Pact in 1990 and the Soviet Union in 1991 demonstrated that Washington's intelligence and policy communities had methodically exaggerated the power and influence of the Soviet bloc, and had invested too much treasure in countering the Soviet threat.

Covert action was a major aspect of our policy for containing the Soviet Union. These actions may have been responsible for a small part of the CIA's budget, but from the Bay of Pigs in 1961 to Nicaragua and Iran-Contra in the 1980s, they have caused the CIA and the United States great embarrassment.

Any discussion of espionage must recognize the incompatibility of covert actions and the American democratic political process. Although the initial use of covert action was designed to buttress democratic elements in postwar Europe, and clearly served U.S. national interests, Hodding Carter III argued in a dissenting opinion to a Twentieth Century Fund report in 1992 that "covert action is by definition outside the ambit of democracy."[5] The United States will probably never abandon covert action entirely, but our democratic principles compel us to define the bounds that should be placed around covert action, to determine what should and should not be attempted, and to insure that there is careful, continuous control over it. As a general principle, we can state that covert action, like military action, should be applied as a last resort, only when vital security interests cannot be achieved in any other way.

It is easy to justify the use of covert action in wartime, but we must decide whether the capacity to conduct covert action in peacetime should continue. The House Intelligence Committee reported in 1996 that, in the clandestine services, "Hundreds of employees on a daily basis are directed to break extremely serious laws in countries around the world in the face of frequently sophisticated efforts by foreign governments to catch them." It went on to state, "A safe estimate is that several hundred times every day (easily 100,000 times a year) DO officers engage in highly illegal activities (according to foreign law) that not only risk political embarrassment to the US but also endanger the freedom if not lives of the participating foreign nations and, more than occasionally, of the clandestine officer himself."[6]

Recent congressional reports on intelligence reform indicate, however, that the foreign-policy establishment is unwilling to tackle serious reform of clandestine operations.[7] Despite recent problems and scandals in the clandestine services, the reports introduce no new thinking on clandestine operations. In fact, they endorse increased spending for covert action and a global presence for the DO.

The need for radical reform is compelling, however. Under three former Directors of Central Intelligence (James Woolsey, Robert Gates, and William Webster), the CIA provided "intelligence" to several Presidents that the Directorate of Operations had obtained from Soviet double agents put in place by the KGB.[8] Instead of acknowledging that they had lost their most important spies in the Soviet Union in 1985 and 1986 and were recruiting only double agents, the DO knowingly provided tainted information to the White House during the final years of the Cold War without informing President Reagan, Bush, or Clinton. When the CIA misleads the President, it is time to start over.

The Origins of Covert Action

The laws authorizing the establishment of the CIA—the National Security Act of 1947 and the Central Intelligence Act of 1949—make no mention of covert action; paramilitary or secret operations; or special operations, all euphemisms for secret warfare. The first Director of Central Intelligence, Rear Admiral Roscoe H. Hillenkoetter, resisted clandestine operations because he was convinced, from his wartime experience, that the Agency could not effectively engage in both information-gathering (the CIA's major mission) and covert action. The CIA's general counsel, moreover, believed that clandestine operations were illegal and that Congress had not intended to grant such authority.

It was the State Department that in 1948 insisted on the need for political and psychological warfare in response to Moscow's efforts to unsettle European governments. The coup d'état in Czechoslovakia and the Berlin blockade led to war jitters in Washington, with George Kennan, the head of policy planning at State, making the first proposal to create a CIA bureau to run covert operations.[9] The bureau, according to Kennan, would report to the Department of Defense in wartime and to the State Department in times of peace.

The CIA began to conduct covert actions in 1947, but received its mandate for such actions only in 1954, from an executive-branch panel charged with examining clandestine operations as an instrument of foreign policy. President Eisenhower was looking for ways to cut the defense budget, and Secretary of State John Foster Dulles believed that covert action was a way to conduct warfare on the cheap. The panel, under the leadership of General Jimmy Doolittle, concluded:

> It is now clear that we are facing an implacable enemy whose avowed objective is world domination by whatever means and at whatever cost. *There are no rules in such a game. Hitherto acceptable norms of human conduct do not apply.* If the United States is to survive, longstanding American concepts of "fair play" must be reconsidered. We must develop effective espionage and counterespionage services and must learn to subvert, sabotage and destroy our enemies by more clever, more sophisticated means than those used against us. *It may become necessary that the American people be made acquainted with, understand and support this fundamentally repugnant philosophy.*[10] [Emphasis added.]

THE ROLE OF COVERT ACTION

The use of covert action, which the Doolittle panel endorsed, is questionable both morally and politically. Covert action, which in the following decades became a staple of U.S. intelligence, contradicts American values and detracts from the message that the United States functions more openly than other nations; this is particularly the case when we take actions abroad that would be neither tolerated nor legal in this country. Most of these actions raise moral and humanitarian questions that tarnish our quest for international stability. Key third-world countries and leaders today, traditionally the primary targets of covert action, are far less vulnerable to U.S. manipulation than in the past, and our allies in Europe and Asia have grown increasingly impatient with the intrusive presence of clandestine officers in U.S. embassies. We must therefore address where, when, and whether covert actions are still necessary. We must ask whether their supposed benefits outweigh their many costs.

Covert "Successes" ...

Covert action has had its presumed successes in terms of meeting U.S. foreign-policy objectives, particularly during the worst days of the Cold War, when the CIA initiated Radio Free Europe and Radio Liberty to bring information to the heavily censored environment of the Soviet Union and Eastern Europe. As noted above, the CIA provided material assistance to Western European trade unions and political parties that were struggling against well-financed communist counterparts. The CIA published books and magazines for the member states of the Warsaw Pact and, with support and guidance from the State Department, developed programs to support democratic regimes in these countries.

More recently, the CIA has mounted covert actions against governments or parties the United States sought to contain, which provided short-term benefits in terms of U.S. policy objectives:

- In 1982, the CIA began a major program, in close coordination with the Vatican, to destabilize the communist government in Poland and support Solidarity, the trade-union movement that opposed that government. The CIA smuggled portable radio and video transmitters to Solidarity through the AFL-CIO's American Institute for Free Labor Development, and created a fund for the legal defense of arrested Solidarity leaders and for fines imposed on illegal publications.[11]
- In 1983, the CIA engaged in unusual cooperation with the government of Iran to crush the underground communist Tudeh party. Using information from a KGB defector, Vladimir Kuzichkin, the United States supplied Iran with information that was used to destroy the Tudeh and the Soviet intelligence network in that country.[12]
- In 1989, after the Chinese crackdown at Tiananmen Square, CIA officers conducted a clandestine rescue of some of the most important pro-democracy leaders. With the support of Ambassador James Lilley, a former CIA Chief of Station in Beijing, the CIA smuggled out students, dissidents, and intellectuals who formed the nucleus of the Chinese democracy movement in exile.[13]
- In 1991, during the Persian Gulf War, the CIA, in conjunction with Polish intelligence, mounted an operation to get U.S. diplomats out of Iraq.[14]
- The CIA's clandestine services also played a major role in the capture of the Cuban revolutionary Che Guevara, the international terrorist Carlos, and Abimael Guzman, the head of Sendero Luminoso (Shining Path) in Peru, which led to a decline in terrorism and insurgency in Latin America and the Middle East.

... And Failures

More often than not, however, covert actions have not been beneficial, and even supposedly short-term policy successes have become long-term failures or liabilities. In Iran, which did not pose any challenge to U.S. national interests in the 1950s, the intense unpopularity of the Shah, whom the CIA had helped return to power in 1953, led to the Islamic revolution of 1979. Major covert actions in Laos and Vietnam failed to alter the results of fighting in Southeast Asia. Interventions in Angola and Mozambique had no effect on conflicts in southern Africa. Covert actions in Nicaragua and El Salvador in the 1980s increased the violence in Central America and brought great embarrassment to the United States.

- In Guatemala, Central America's most brutal regime was installed in 1954 with the help of a CIA-backed coup. The country was dominated by its repressive military for the next forty years. Government documents show that in 1990, Colonel Julio Roberto Alpirez, a CIA informer, was involved in the cover-up of the murder of Michael Devine, an American citizen, and that in 1992 he helped cover up the murder of Efrain Bamaca Velasquez, a Guatemalan insurgent who was married to Jennifer Harbury, an American citizen. Such episodes undermined the credibility of the CIA and raised questions about its judgment and objectivity.

    In 1997 the CIA released a small batch of records on the 1954 military coup in Guatemala, but it has declassified practically nothing on the Guatemalan security forces, which have killed an estimated 200,000 Guatemalans since the coup. The Agency trained and supported some of these forces, along with similarly abusive internal security organizations in Nicaragua, Haiti, Honduras, and El Salvador.[15]
- In Honduras, government officials have been risking their lives to prosecute some two dozen military men involved in a death squad that killed at least 184 people in the early 1980s. The death squad grew out of a collaboration between the CIA and the Honduran military. Recently declassified documents reveal that "CIA reporting linked Honduran military personnel to death squad activities."[16] These documents confirm that the CIA knew that the Honduran military committed repeated and systematic human-rights abuses in the 1980s, but that the Agency continued to collaborate with its Honduran partners and misled Congress about the abuses.
- Robert M. Gates, former Director of Central Intelligence, termed support to the mujahideen fighting the Soviets in Afghanistan the CIA's "greatest success," but today Afghanistan is a country of death and misery; weaponry supplied to the mujahideen by the Agency is fueling conflicts in Bosnia and the Sudan; and rebels trained by the CIA were involved in terrorist actions in New York.[17] Agency-backed mujahideen equipped with surface-to-air Stinger missiles may have highlighted the policy dilemma faced by Soviet leaders, but the Kremlin's decision to withdraw from Afghanistan was made *before* the arrival of the missiles.[18] In 1994, the CIA had to begin a $65 million covert effort to buy back the Stingers, which in the hands of the mujahideen had become a long-term security and terrorist problem, just as many U.S. officials had warned in the 1980s.[19] Ironically, the Russians are now supporting antigovernment rebel factions in Afghanistan that were formerly supported by the CIA.

The CIA's role in Afghanistan still remains largely unknown. Under the direction of the Reagan administration, the Agency spent $3 billion in the 1980s to train and fund seven Afghan resistance groups—all venomously anti-American—that have formed the core of an international network of highly disciplined and effective Islamic militants. Much of the CIA weaponry went to the fundamentalist Gulbuddin Hekmatyar, one of the most anti-Western of the resistance leaders and now the Prime Minister of Afghanistan. Hekmatyar's allies included Sheikh Omar Abdul-Rahman, who was imprisoned in New York for seditious conspiracy to wage a "war of urban terrorism against the United States," and Muhammad Shawqi Islarnbouli, the older brother of the assassin of Anwar Sadat. The terrorist network has targeted Saudi Arabia, Egypt, and Pakistan—Washington's most pivotal Islamic allies in the region—and has claimed responsibility for the first terrorist attack in Saudi Arabia and some of the worst attacks in Pakistan. Osman bin Laden, who has led terrorist attacks against U.S. installations in the Persian Gulf and North Africa, received CIA support in Afghanistan during the 1980s.

- Many of the most spectacular CIA failures occurred during the Kennedy administration, when the Agency misread or ignored the likely effects of its actions on target nations. The nadir of the CIA's covert actions was the Bay of Pigs invasion of Cuba in 1961, when high-level ignorance of Fidel Castro's popularity led the CIA to launch its ill-fated paramilitary operation. Even if the operation had gone flawlessly, it would not have dislodged Castro. In the wake of the failure of the Bay of Pigs, Cubans who received CIA training conducted their own freelance anti-Castro operations. Several Cubans, initially trained by the CIA for covert action against their home country, were involved in the break-in of the Democratic National Committee's offices in the Watergate complex ten years later.[20]

In 1998, after thirty-six years of secrecy, the CIA finally released its Inspector General's report on the Bay of Pigs, a startling and brutally honest inquest on the Agency's greatest fiasco.[21] The Agency had destroyed nearly every copy of the report except for those of the Director of Central Intelligence, the Inspector General, and the Records Center. The report put the blame squarely on what it described as "arrogance, ignorance, and incompetence" within the CIA, found that the Agency was shot through with deadly self-deception, and described its secret operations as "ludicrous or tragic or both." Since plausible denial is the sine qua non of covert action, it is noteworthy that the report concluded that in the Bay of Pigs "plausible denial was a pathetic illusion," and argued that in the future paramilitary operations should be conducted by the Pentagon.

- Revelations of assassination plots in Cuba, the Congo, the Dominican Republic, and perhaps Vietnam in the early 1960s—again at the direction of the Kennedy administration—finally led to a ban on CIA political assassinations in the mid-1970s. Nevertheless, covert action in Ethiopia in the early 1980s, according to Ambassador David Korn, led to the deaths of many people, including some who were entirely innocent and extraneous to the CIA's attempt to overthrow the government of Haile-Maryam Mengistu. Honduran commandos trained by the CIA killed and tortured people suspected of helping Salvadoran rebels, and the recipients of CIA antiterrorist training in Lebanon set off a car bomb that killed eighty innocent people in Beirut. In 1984, the CIA showed how seriously it took the ban on assassination when it produced a manual for the Contras that discussed "neutralizing" officials in Nicaragua.[22]
- Recently, Congress secretly rejected a CIA plan for a new covert operation against Iraq's Saddam Hussein. Congressional leaders reportedly complained that the CIA was trying to "slip in a highly sensitive program at the last minute without seeking reviews by the House and Senate intelligence committees." The last covert action against Baghdad took place in 1996, when President Hussein destroyed the two most important CIA programs in Iraq, which were attempting to destabilize his regime, and ordered the execution of 100 army officers and other opponents connected with the Iraqi National Accord, a CIA-backed group.
- In addition to covert action against regimes identified by the Agency as dictatorial or subversive, there have been actions against identified neutrals such as Indonesia and democracies such as Chile. In the late 1950s, the CIA, reflecting the Eisenhower administration's intolerance of neutrals, tried to unseat President Sukarno of Indonesia. In 1970 it tried to prevent the election of Salvador Allende, a leftist, as President of Chile. After Allende's election, the CIA moved to subvert his government. There was no justification for these actions in terms of American national security or national interest. Henry Kissinger, President Nixon's National Security Adviser, described Chile as a "dagger pointed at the heart of Antarctica," but he did not "see why the United States should stand by and let Chile go communist merely due to the stupidity of its own people."[23]

Kissinger's efforts against Chile were particularly outrageous, involving bribes to members of the Chilean congress in order to allow the Christian Democratic President, Eduardo Frei, to run again in violation of the country's constitution. Kissinger authorized covert propaganda to convince Chile's congress that the economy would be ruined if Frei were not reelected. He also authorized the introduction of two dozen CIA-sponsored journalists from around the world to produce stories hostile to Salvador Allende, the Marxist candidate. The instructions from the President and the National Security Adviser to Richard Helms, head of the CIA, were explicit, allowing Helms to say that "If I ever carried a marshal's baton in my knapsack out of the Oval Office, it was that day."[24]

Frei and his Christian Democrats wanted no part of Kissinger's chicanery; they were less afraid of an Allende victory than of a covert scheme to trample the electoral process. The commander in chief of the Chilean army, General Rene Schneider, also wanted to protect Chile's democracy and was opposed to the idea of a military coup, which Kissinger favored. But Kissinger had no qualms about interfering with the democratic electoral process. The CIA provided money and machine guns to right-wing renegades who were to kidnap and kill Schneider, but others got to Schneider before the CIA-sponsored groups. The general was killed two days before the congress voted Allende into office.

- Far too often, paramilitary aid from the CIA has carried with it substantial danger for the recipients. The CIA established covert relations with the Kurds in northern Iraq, the Bay of Pigs invaders, and the Meo tribesmen in Laos, only to abandon them when U.S. policy changed. The CIA encouraged rebellion among Hungarian freedom fighters in the mid-1950s, but the United States did nothing when the Hungarian revolution began in 1956. The U.S.-backed Contra war against Nicaragua's Sandinista government and the CIA support for UNITA in Angola in the 1980s raised the level of violence in both nations, with CIA stations in Central America and Africa keeping sensitive information from the Congress and U.S. Ambassadors.

We now know, moreover (see Chapter 7), that the CIA was aware of the efforts of the Nicaraguan Contras to traffic in drugs in the United States. According to recently released CIA studies, the CIA worked with two dozen Nicaraguan rebels and their supporters during the 1980s despite knowledge that they were trafficking in drugs.[25] As early as 1981, the CIA knew that the Contra leadership "had decided to engage in drug trafficking to the United States to raise funds for its activities." The leader of the Contra group was Enrique Bermudez, whom the CIA had picked to run the military operations of the Contra organization. Director of Central Intelligence Casey made no attempt to report these activities to the Justice Department, as required by law, and even obtained a waiver from Justice to provide a "legal basis" for his illegalities. Casey also ordered the Drug Enforcement Administration "not to make any inquiries" regarding the

airfield in El Salvador that was being used as an arms-for-drugs transshipment point.
- Support for covert action in one area often compromises U.S. strategic objectives in other areas. The need to channel arms to the mujahideen through Pakistan, for example, meant that the United States had to ignore Pakistan's nuclear-weapons program in the 1980s. During that period, Casey and Gates kept sensitive intelligence dealing with Pakistan's nuclear program from the Senate Select Committee on Intelligence. The need to channel arms to UNITA forces in Angola through Zaire meant ignoring the brutality of Mobutu's rule in Zaire. The CIA funded the creation of military units in Central America, such as Battalion 316 in Honduras, that were responsible for the deaths of innocent civilians. Various administrations looked away from human-rights violations and drug trafficking in Pakistan and South Africa because of covert programs that required support from U.S. proxies.
- Finally, we have learned that for three years, the United States and the Central Intelligence Agency used the cover of the United Nations and the United Nations Special Commission (UNSCOM) to conduct a secret operation spying on Iraqi military communications. The United States had consistently denied Iraqi charges that it was exploiting U.N. inspections for the purposes of American espionage. The U.N. effort, which did not authorize or benefit from U.S. surveillance, was designed to gather information on Iraq's suspected nuclear, chemical, and biological weapons. But the CIA program was not aimed at Iraq's weapons of mass destruction. Instead, CIA operatives were trying to penetrate Iraqi communications in order to bring down the regime of Saddam Hussein, a covert action sanctioned by the U.S. Congress in the Iraqi Liberation Act. This shabby exploitation of the United Nations doomed further inspection efforts in Iraq and undercut the credibility of multilateral inspection efforts around the world. The CIA also made a liar out of the White House and a truth-teller out of Saddam Hussein.[26]

REFORM OF COVERT ACTION

If the Cold War and the Soviet threat generated the rules that governed the use of covert action, then the end of the Cold War and the dissolution of the Soviet Union make the need for clandestine operations less demonstrable and demand a reexamination of every aspect of such operations. It is not enough to suggest—as defenders of covert action have suggested—that the world remains a dangerous place and that the President needs an option short of military action when diplomacy alone cannot do the job.

Most problems that could be considered candidates for covert action could be addressed openly by unilateral means or cooperatively through international measures. Nuclear-proliferation problems created by missile programs in Iraq and North Korea in the 1990s led to calls for covert action, but in both cases overt multilateral activity, with the United States playing the pivotal role, contributed to denuclearization.

Covert action could be radically reduced with no compromise of U.S. national security. CIA propaganda has had little effect on foreign audiences and should end immediately. Covert efforts to influence foreign elections or political parties should stop. It was wrong for the Bush administration to authorize a covert operation to return Miami-based Contra leaders to Nicaragua for the presidential elections in 1990, after the Congress had approved an expenditure of $9 million on the election through the National Endowment for Democracy and had banned covert support to Violetta Chamorro's National Opposition Union.

The boilerplate language of the report of the Brown Commission on intelligence reform in 1996 argued that covert operations should take place only when "essential" and where the reason for secrecy is "compelling." But in my view, most covert operations are "operations for operations' sake," undertaken with no careful reckoning of the result beforehand. There is no absolute political and ethical test for covert action, but Cyrus Vance, the former Secretary of State, articulated a good standard two decades ago when he recommended covert intervention only when "absolutely essential to the national security" of the United States and when "no other means" would do.[27]

In 1998, rather than cutting back on such actions, Director of Central Intelligence George Tenet lobbied Congress for an increased capability to conduct clandestine operations, and the 1998 Senate Intelligence Committee favored the hiring of additional clandestine officers. The recommendation of the chairman of the House Intelligence Committee was to create a separate organization for espionage operations. This would lead to a renewed mandate for covert action, with increased funding and more personnel. Cutting the clandestine services adrift, moreover, would result in the concentration in one organization of most of those officers who have been exposed to the highest ethical risks. As a high-level Agency official argued nearly twenty years ago, "backsliding would be a great temptation, managerial control an administrator's nightmare."[28]

THE ROLE OF CLANDESTINE COLLECTION

There is a tendency in the U.S. government to overemphasize the value of clandestine collection, honoring the myth that whereas open collection can track the capabilities of foreign nations, only covert collection can determine the intentions of foreign leaders. This theory is flawed; it is virtually impossible to divine the intentions of a foreign leader unless the leader himself is the source. CIA sources failed to decipher Nikita Khrushchev's intentions toward Hungary in 1956, Leonid Brezhnev's toward Czechoslovakia in 1968, Anwar Sadat's toward Israel in 1973, Menachem Begin's toward Lebanon in 1982, or Saddam Hussein's toward Kuwait in 1990. The CIA has had a mediocre record of collecting intelligence on the possibilities of war and coups. On balance, the intelligence community is good at finding things, counting things, and describing things, but intentions are a more elusive quarry.

The product of human intelligence, or HUMINT, is, in fact, not all that impressive. Robert M. Gates conceded in his memoirs that the Agency "never recruited a spy who gave us unique political information from inside the Kremlin." The CIA has had little success collecting intelligence in closed societies, and much of its clandestine reporting on the Soviet Union, Eastern Europe, and Cuba came from double agents. Intelligence has tended to be collected as an end in itself, when collection should be focused on a very few high-priority targets. An exception to this pattern was the clandestine gathering of intelligence on Soviet weapons systems during the Cold War, which enabled all U.S. administrations to have a great deal of warning time on the procurement and deployment of new weaponry in the Soviet Union.

Far more "intelligence," particularly in political and economic areas, can be gleaned from the reporting of U.S. Ambassadors and embassy officers, who build close personal and open relations with influential government officials. Clandestine collection of intelligence has been marginal in closed societies and often inferior to information from Foreign Service officers, military attachés, and journalists in open societies. It would be more cost-effective to shift resources devoted to clandestine collection to the State Department and other agencies, which have had to reduce their personnel in the field because of budget limitations. The budgets of the State Department and the United States Information Agency have been steadily reduced in recent years, while that of the intelligence community has been increased. Today, the State Department's budget is smaller than the CIA's.

In making the CIA's clandestine collection the "principal or sole source of information" in closed societies, there is a tendency to overlook the Directorate of Operations' past reliance on double agents for much of its intelligence. Robert Kimmett, former Ambassador to Germany, and others have charged, moreover, that CIA operators have used Agency funds to recruit embassy sources, thus making the United States pay for intelligence that is usually available without cost.[29] (See Chapter 4.) U.S. Information Agency officers often lose valuable sources abroad to the blandishments of

CIA operators. Robert White, formerly the U.S. Ambassador to El Salvador and Paraguay, has suggested that the CIA could close its stations in many parts of the world—Central America, for example—and allow career diplomats to fulfill Washington's intelligence requirements.[30] (See Chapter 3.) Intelligence collection would not suffer, and diplomatic influence would be enhanced. Foreign-service and clandestine officers have acknowledged that the CIA presence in Central and South America is aimed primarily at the drug market, and have questioned the use of CIA involvement.

The report of the Council on Foreign Relations in 1996 took a step backward with its implicit endorsement of expanded use of CIA cover to include journalists, clergy, and Peace Corps volunteers. This suggested misuse of the Peace Corps would destroy its integrity as a "nonpolitical" humanitarian organization, and would greatly increase the danger to its volunteers. The House Intelligence Committee, in its 1996 report, also recommended that the clandestine services apply journalistic cover to their operators abroad. There is no justification for the use of spies posing as reporters or the employment of bona fide reporters for intelligence missions, practices that developed during the Cold War. Both practices should be banned. The press has constitutional protection because it is the chronicler of and check on the government, not its instrument. Unfortunately, recent CIA directors have insisted that the Agency have the option of using journalists in sensitive clandestine operations.

The recent reports on intelligence reform, particularly the Council on Foreign Relations' *Making Intelligence Smarter,* in 1996, defended U.S. association with "unsavory" characters, including "some who have committed crimes," when we get something in return. This, however, has rarely been the case, particularly if the long-term consequences of these associations are considered.[31] In any event, what little value we may have received in the short term hardly justifies being a party to criminal activity. One of the first associations the CIA made in Contra recruitment and training was with the Argentine generals, whose repressive rule was overlooked. Another channel to Contra support was Panama's General Manuel Noriega, who had for some time been on the CIA payroll and now resides in a Florida prison. CIA sources have been involved in the cover-up of criminality throughout Central America.

Clandestine collection, as Roger Hilsman points out, is still needed against so-called rogue states as well as against nations that are unfriendly to the United States. It is also needed to track the proliferation of weapons of mass destruction, military developments in Russia and China, and international terrorism. Unfortunately, the CIA often ignores these priorities and, as in the 1998 bombing of the pharmaceutical plant in Sudan, rushes to judgment on the basis of incomplete collection and analysis.

The failure of the CIA in 1998 to monitor Indian nuclear testing is a classic example of the weakness and occasional irrelevance of the clandestine mission abroad. That failure exposed, moreover, the major deficiency evident at the CIA for the past several years: a fundamental misunderstanding of the proper interaction between intelligence and policymaking. Since U.S. policymakers did not expect nuclear testing by India, U.S. intelligence officials assigned a low priority to problems on the subcontinent. In other words, policymakers determined the agenda for all types of intelligence in South Asia, including clandestine collection by the CIA and imagery analysis at the National Imagery and Mapping Agency. The Department of Defense, which is now responsible for all analysis of satellite imagery, assigned a low priority to all political-military issues in South Asia. As a result, there was insufficient satellite or clandestine collection on a country about to break the ban on nuclear proliferation.

REFORM OF CLANDESTINE COLLECTION

In the 1990s, the CIA had begun to focus on high-priority targets. The Agency closed many stations and bases around the world, particularly in Africa, and it seemed likely that financial constraints would lead to additional closures. The size of CIA stations also was reduced, and there are currently fewer case officers than at any time in the past twenty years. Along with these moves to lower the number of stations and case officers, the clandestine service concentrated the scope of its collection on targets considered central to U.S. interests and objectives.

The director of the Agency, George Tenet, has taken steps to reverse this trend, however. He is promoting the notion of a global presence for the CIA rather than focusing collection on high-priority targets such as proliferation, terrorism, and military technology. The House and Senate Intelligence Committees have unwisely endorsed Tenet's global mission for the CIA, and do not support additional cutbacks. Tenet has taken steps to exploit the current mood of the Congress.

THE POLITICIZATION OF INTELLIGENCE

In 1996, the Brown Commission endorsed a step that will increase policy advocacy within the CIA—a new "partnership" between the directorate that carries out espionage and covert action, the DO, and the one that analyzes intelligence, the Directorate of Intelligence, or DI. Although the idea appears to be dead for now, any merger of the DO and the DI would give CIA clandestine reports a privileged status over other reporting and make it possible for the clandestine service to influence the analytical product of the CIA. The CIA already is putting too much emphasis on current reporting and ignoring the role of long-term research and strategic intelligence; eventually the role of intelligence analysis could be subsumed into support for clandestine collection. A merger would also make it more difficult to recruit senior, outside experts from the academic community.

In 1994, James Woolsey, President Clinton's first DCI, tried to locate the DO and the DI in the same offices as the first step toward an actual merger, without any study of the implications of such an arrangement. His successor, John Deutch, unfortunately endorsed the idea of a merger and even named a Deputy Director for Operations who had misused clandestine reporting in the CIA's specious intelligence assessments. Woolsey's action came in response to the embarrassing Ames affair in the 1980s, in which a trusted, high-level CIA official turned over damaging intelligence to the Soviets, including the names of at least ten U.S.-paid foreign agents, who were subsequently executed. But the director failed to take into account that the Agency's two major divisions cannot be reformed in the same way. The operations wing (because of covert action) is deeply involved in policy; it relies on secrecy and hierarchy, and shares information on a need-to-know basis. Its culture is a military one, relying on a chain-of-command structure and paramilitary training.

In contrast, the intelligence wing must have no policy axes to grind; its credibility rests on that fact. It needs a free exchange of information and a great variety of analytical input. The Brown Commission incorrectly concluded, for example, that most information on the former Soviet Union was "secret and could best be obtained, analyzed, and reported by the Intelligence Community."[32] In fact, most political and economic intelligence was gleaned from open sources, and should have been scrutinized by a wide community of scholars and analysts. The operational culture's preoccupation with secrecy is a threat to the openness required by the culture of intelligence analysis.

In approving the merger of the operational and intelligence directorates, the Brown Commission ignored the advice of a key witness, Admiral Bobby Inman, former head of the National Security Agency and former Deputy Director of Central Intelligence. Inman warned against merging collection and analysis of intelligence in a single agency, explaining that analysts in such agencies are under a great deal of pressure to "protect, but not to challenge the information" that these agencies collect. William Casey demonstrated this problem when he was CIA director in the 1980s, integrating operations and intelligence so that he could influence analysis to support his preferred policies toward Central America and Southwest Asia, particularly Afghanistan. Robert M. Gates stated that he watched Casey "on issue after issue sit in meetings and present intelligence framed in terms of the policy he wanted pursued."[33] Mixing operatives and analysts invites politicization because operatives have a tie to the policy process that will "infect" analysis. Having the CIA director sit in the President's Cabinet, moreover, provides an opportunity to politicize sensitive clandestine collection.

In the recent past, any operational role for the CIA on behalf of policy has led to the politicization of the collection and analysis of intelligence. This was true during the 1980s, when the CIA was heavily involved in covert action on behalf of the National Security Council in Afghanistan and Central America and, as a result, collected intelligence that was supportive of NSC policy. The CIA then crafted analysis to mislead both policymakers and Congress.

The 1998 peace agreement in the Middle East (the Wye Accord) states that the CIA is to verify key aspects of the agreement. In other words, it will be collecting intelligence on both Israeli and Palestinian actions. Will this intelligence, which will be critical to the success of U.S. policy, be politicized or skewed before it reaches the White House? Will the CIA be able to conduct electronic surveillance of Yasir Arafat's Palestinian Authority while trying to establish itself as an independent arbiter of the peace agreement? Has the CIA been given a sensitive diplomatic mission to verify a peace accord because of the lack of funding at the State Department?

As recently as 1998, Representative Porter J. Goss, chairman of the House Intelligence Committee and a former clandestine-services officer, argued for increased funds and greater hiring in the CIA's Directorate of Operations. "The cupboard is nearly bare in the area of human intelligence," Goss said, and it is time for the CIA to be "bold and imaginative" in finding new ways to deal with the leaders of rogue countries, such as Iraq's Saddam Hussein.[34] Instead of urging the examination of overt, diplomatic, and multilateral ways to address new problems in the post–Cold War era, Goss wants "more arrows in the quiver" of covert action, particularly in the areas of "cyberspace" and "mind management," in order to create doubt among an enemy's leader's supporters.

With the two inconclusive government reports in 1996 and its subsequent inaction, the United States has squandered its best opportunity for intelligence reform since the post-Watergate investigations of the CIA in 1975. Those investigations, known as the Pike and Church Committees, produced a great deal of information on the intelligence community but few reforms, other than the creation of new standing committees on intelligence. The Pike Committee, named after Representative Otis G. Pike, who headed the House Select Committee on Intelligence, suggested in its 1975 report that the CIA was propping up dictatorships, listed CIA failings, and claimed that Congress had been misled about intelligence costs. The Church Committee, named after Senator Frank Church, who headed the Select Committee to Study Governmental Operations with Respect to Intelligence Activities, applauded, in its 1976 report, the presidential ban on assassinations, and recommended legislation to prohibit the subversion of democratic governments, to outlaw clandestine support for repressive regimes that ignore human rights, and to prevent federal agencies from electronically spying on U.S. citizens.

The House overwhelmingly rejected the work of the Pike Committee, and the Senate rejected most of the recommendations of the Church Committee. Thirteen years after these investigations, in the wake of Iran-Contra, the joint congressional investigating committee concluded that the existing oversight laws were adequate and that the system had worked. Once again, the Congress appropriated additional funds for the intelligence community and the Central Intelligence Agency. The more things change, the more they stay the same.

What Needs to Be Done?

The end of the Cold War and the stunning array of CIA failures have produced another opportunity for reform, but the 1996 report of the Brown Commission, in particular, evaded the challenge. The commission favored spending more on intelligence, not less, and more espionage and covert action. The House Intelligence Committee, in its 1996 report, also recommended an increase in intelligence spending. The former chairman of the House committee, Representative Larry Combest, even favored removing the DO from the CIA and combining it with the Pentagon's defense HUMINT service to create a comprehensive organization for espionage. The effect would be an even stronger clandestine capability, less susceptible to congressional oversight.

The fact that the CIA passed to several U.S. Presidents, including President Clinton, information obtained from Soviet double agents is reason enough for a major shake-up of the clandestine corps. Recent spy scandals involving CIA efforts to collect economic intelligence in France, Germany, India, Italy, and Japan, which contain some of the largest CIA stations, raised additional questions about the judgment and goals of the Directorate of Operations. The CIA supported military organizations throughout Central America despite the long history of their human-rights abuses and the limited value of clandestine collection in the region. Covert action to influence the course of events in a foreign country without our role being known undermines our own standards of openness and honesty and should be authorized only in times of national peril.

The end of the CIA's original mission as an instrument to combat the Soviet Union presents Congress with an opportunity to reexamine the need for clandestine operations. The late Richard Bissell, onetime head of covert operations and the planner of the Bay of Pigs operation, conceded that there "have been many short-term tactical victories" in covert action, but "very few lasting successes." It is time to transfer the CIA's paramilitary capabilities to the Defense Department, to end covert military intervention, and to discuss a ban on covert action in peacetime. If the capability for covert military action is deemed essential for U.S. security in wartime, that function would be best located within the Pentagon, not the CIA. Overt diplomatic and military means—both unilateral and multilateral—should be sufficient when there are threats to our national interest in peacetime. Covert political action is rarely necessary, and covert action in peacetime can only compromise our principles as a constitutional democracy.

We should continue the clandestine collection of intelligence against the proliferation of nuclear, chemical, and biological weapons; international terrorism; and regimes that threaten regional stability. But this can be done with a far smaller operational presence in selected capitals, where CIA stations should focus on key issues. Unfortunately, recent CIA directors have focused clandestine collection on support for military operations, which marginalizes the role of the CIA in strategic and political intelligence. The House Intelligence Committee even favors naming a flag-rank officer Deputy Director for Clandestine Intelligence, which would enhance the militarization of the CIA.

The failure to address problems created by covert action has been compounded by the failure to address the problem of oversight. The intelligence oversight committees and their staffs are small in comparison with the community they monitor; they must be strengthened. The National Reconnaissance Office managed to misplace $4 billion because of obsessive secrecy and weak oversight, then failed to return the unspent funds to the Congress, as is required by law. Oversight of clandestine activities is deficient and should be strengthened. Information about intelligence operations in other countries is often exempt from ambassadorial scrutiny; this must be corrected. (See Chapters 8 to 10.)

Nevertheless, the presidential commission on intelligence reform (the Brown Commission) insisted in 1996 that the Senate and House committees have provided "rigorous and intensive oversight," and the House study introduced a new (and shocking) role for congressional oversight: that of "advocate." As Steven Aftergood argued in *Secrecy and Government Bulletin,* any "advocacy role is fundamentally incompatible with credible oversight and is in fact the reduction to the absurd of the oversight mission."[35]

For too long, the Intelligence Committees have been advocates for the CIA—particularly for the clandestine world of spies and covert operations. Thus Congress has failed to make the CIA accountable for its transgressions, and Iran-Contra demonstrated that no form of accountability was sufficient to monitor ideological zealots in the CIA and the National Security Council. A presidential pardon in 1992 for key CIA operatives involved in Iran-Contra means that we will never know the extent of CIA perfidy in presidential maneuverings related to that scandal.

In a democracy, where laws are derived from broad principles of right and wrong and where those principles are protected by agreed procedures, it is not in the interest of the state to flout those procedures at home, or to associate overseas with the enemies of this nation's founding ideals. The Intelligence Committees, at the very least, should take a hard look at the British system, where espionage and clandestine collection are conducted by an organization that reports to the highest levels of the British Foreign Office. Research and analysis is conducted by an organization that is separate from both the Foreign Office and the military. Radical reform would allow the CIA to return to President Truman's original conception of it as an independent and objective interpreter of foreign events.

NOTES

1. Robert M. Gates, *From the Shadows: The Ultimate Insider's Story of Five Presidents and How They Won the Cold War* (New York: Simon and Schuster, 1996), p. 32.

2. See Kathryn S. Olmsted, *Challenging the Secret Government: The Post-Watergate Investigations of the CIA and FBI* (Chapel Hill, NC: University of North Carolina Press, 1996).

3. Interviews with CIA officials, Washington, DC, December 11, 1998.

4. *Hearings Before the Select Committee on Intelligence of the United States Senate: Nomination of Robert M. Gates* (Washington, DC: Government Printing Office, 1991), p. 38.

5. Twentieth Century Fund, *The Need to Know: The Report of the Twentieth Century Fund Task Force on Covert Action and American Democracy* (New York: Twentieth Century Fund Press, 1992), pp. 21–23.

6. House Permanent Select Committee on Intelligence, *IC 21: The Intelligence Community in the 21st Century* (Washington, DC: Government Printing Office, April 9, 1996), p. 205.

7. Commission on the Roles and Capabilities of the United States Intelligence Community, *Preparing for the 21st Century: An Appraisal of U.S. Intelligence* (Washington, DC: Government Printing Office, March 1, 1996); Council on Foreign Relations, *Making Intelligence Smarter: The Future of U.S. Intelligence* (New York: Council on Foreign Relations, 1996; *IC 21*.

days with a receipt from any Barnes & Noble store.
Store Credit issued for new and unread books and unopened music after 30 days or without a sales receipt. Credit issued at <u>lowest sale price</u>.
We gladly accept returns of new and unread books and unopened music from bn.com with a bn.com receipt for store credit at the bn.com price.

Full refund issued for new and unread books and unopened music within 30 days with a receipt from any Barnes & Noble store.
Store Credit issued for new and unread books and unopened music after 30 days or without a sales receipt. Credit issued at <u>lowest sale price</u>.
We gladly accept returns of new and unread books and unopened music from bn.com with a bn.com receipt for store credit at the bn.com price.

Full refund issued for new and unread books and unopened music within 30 days with a receipt from any Barnes & Noble store.
Store Credit issued for new and unread books and unopened music after 30 days or without a sales receipt. Credit issued at <u>lowest sale price</u>.
We gladly accept returns of new and unread books and unopened music from

8. See Melvin A. Goodman, "Ending the CIA's Cold War Legacy," *Foreign Policy*, No. 106, Spring 1997.

9. See Evan Thomas, *The Very Best Men: Four Who Dared: The Early Years of the CIA* (New York: Simon and Schuster, 1995).

10. "Report of the Special Study Group (Doolittle Committee) on the Covert Activities of the Central Intelligence Agency" (September 30, 1954), in William M. Leary (ed.), *The Central Intelligence Agency: History and Documents* (Tuscaloosa, AL: University of Alabama Press, 1984), p. 144.

11. Raymond Garthoff, *The Great Transition: American-Soviet Relations and the End of the Cold War* (Washington, DC: The Brookings Institution, 1994), pp. 32–33.

12. Bob Woodward, "CIA Favors Khomeini, Exiles; Sources Say Agency Gave Them List of KGB Agents," *Washington Post*, November 19, 1986, pp. 1, 28.

13. Mark Perry, *Eclipse: The Last Days of the CIA* (New York: William Morrow, 1992), p. 246.

14. John Prados, *Presidents' Secret Wars: CIA and Pentagon Covert Operations from World War II Through the Persian Gulf* (Chicago: Elephant Paperbacks, 1996), p. 486.

15. *New York Times*, September 13, 1998, p. 7; February 26, 1999.

16. *New York Times*, February 22, 1998, p. 6.

17. Gates, *From the Shadows*, p. 51.

18. See Diego Cordovez and Selig S. Harrison, *Out of Afghanistan: The Inside Story of the Soviet Withdrawal* (New York: Oxford University Press, 1995); Georgy Korniyenko and Sergei Akhromeyev, *Through the Eyes of a Marshal and a Diplomat: A Critical Look at the Foreign Policy of the USSR Before and After 1985* (Moscow: International Relations Publishing House, 1985); Carolyn M. Ekedahl and Melvin A. Goodman, *The Wars of Eduard Shevardnadze* (University Park, PA: Penn State Press, 1996).

19. Prados, *Presidents' Secret Wars*, p. 48.

20. Pat M. Holt, *Secret Intelligence and Public Policy: A Dilemma of Democracy* (Washington, DC: Congressional Quarterly Press, 1995), p. 162.

21. Tim Weiner, "CIA Bares its Bungling in Report on Bay of Pigs Invasion," *New York Times*, February 22, 1998, p. 6.

22. Gates, *From the Shadows*, p. 310.

23. Gregory F. Treverton, *Covert Action* (New York: Basic Books, 1987), p. 11.

24. Walter Isaacson, *Kissinger: A Biography* (New York: Simon and Schuster, 1992), p. 290.

25. James Rison, "CIA Says It Used Nicaraguan Rebels Accused of Drug Tie," *New York Times*, July 17, 1998, p. 1.

26. *Washington Post*, March 2, 1999, p. 1.

27. Testimony Before the Senate Select Committee on Intelligence Activities, December 5, 1975; Executive Order 12036 (Washington, DC: Government Printing Office), 1978.

28. Gates, *From the Shadows*, p. 77.

29. Testimony of the American Foreign Service Association to the Brown Commission on the Roles and Capabilities of the Intelligence Community in the Post–Cold War Environment, January 19, 1996, p. 2.

30. Robert E. White, "Call Off the Spies," *Washington Post,* February 7, 1996, p. 27.

31. *IC 21,* p. 193.

32. *Preparing for the 21st Century,* p. 134.

33. *New York Times,* June 15, 1996, p. 11.

34. *Washington Post,* May 10, 1998, p. 4.

35. *Secrecy and Government Bulletin,* No. 56, March 1996, p. 2.

Robert E. White

3 Too Many Spies, Too Little Intelligence

WHAT THE CENTRAL INTELLIGENCE AGENCY has bequeathed to our relations with Central America and the Caribbean is a string of embarrassing failures against inconsequential targets. From the overthrow of the government of Guatemala to the Iran-Contra fiasco of the 1980s, the CIA not only violated solemn treaties but allied us with the most violent, reactionary elements of Latin American society. In carrying out these operations, the CIA subverted American values at home as well as abroad.

During the Cold War, the United States demanded the subordination of Central America's political and social priorities to our supposed national-security requirements. With a wink and a nod from the United States, the militaries of Central America arrogated to themselves the right to decide who could and could not participate in the political life of the country. In the name of anticommunism, U.S.-supported and -trained military establishments suppressed democracy, free speech, and human rights in Guatemala, El Salvador, Nicaragua, Honduras, and Panama. The torture and assassination of democratic leaders, including presidential candidates, journalists, priests, and union officials, became commonplace.

Such policies were pursued without the adequate safeguards of accountability that characterize a democracy. As Senator Daniel Patrick Moynihan wrote in his fine book *On the Law of Nations*, "As the United States became more committed to the advancement of democratic values in the world at large, it came more and more to do so by means of covert strategies, concealed from the world and not least from the American public. This is not difficult to explain; it is difficult to defend. It costs too much, it achieves too little; and it gives power to presidents to do things that come to seem merely extralegal, rather than illegal. Not lawless, simply above the law. The intelligence community cannot help but to make presidents feel this is what they are there for."[1]

Covert action works at cross-purposes with the open diplomacy of the Foreign Service. American embassies and consulates around the world exist to safeguard American interests in a manner consistent with national honor. Diplomacy divorced from ethical standards and international law is simply bad statecraft. It betrays the values of the United States and undercuts the relationships of trust and confidence with foreign governments so necessary to the achievement of our national objectives.

Guatemala

If we had to choose one time and place where U.S. policy toward Latin America went wrong, the date would be 1954 and the place would be Guatemala. Confronted with an elected government less than subservient to U.S. pressure, the Eisenhower administration gave the CIA a green light to overthrow the constitutional order and install a military government, thereby igniting widespread rebellion.

In March 1999, the Guatemalan Commission on Historical Verification issued a report that not only found the Guatemalan military responsible for mass murder and genocide but stated flatly that the "government of the United States through various agencies including the CIA provided direct and indirect support for some state operations." At the presentation of the report, the commission's coordinator, Christian Tomuschat, said its investigations revealed that "until the mid-1980s there were strong pressures from the United States to keep in place the archaic and unjust socioeconomic structures of the country," and that the United States and the CIA supported "illegal state operations."[2]

The Guatemalan commission's report makes it clear that the United States was backing not a war but a campaign of state terror. With direct orders from the government's highest echelons and the military high command, soldiers carried out a scorched-earth policy, burning Mayan villages and throwing living victims into common burial pits. Declassified documents reveal that the U.S. government knew of these acts of genocide and nevertheless not only continued but increased its assistance to the Guatemalan military.

Although the Defense Intelligence Agency and the Department of State must share responsibility for this policy, Guatemala's Commission on Historical Verification was right to single out the CIA. Between 1965 and 1981, I served in our embassies in Honduras, Nicaragua, and El Salvador. I watched as the CIA recruited dozens of paid informants from the lunatic right-wing fringe of Central American society. These ideologues regarded labor-union leaders threatening a strike or student activists protesting the closing of another newspaper as agents of subversion.

I watched as CIA reports to Washington characterized as communist sympathizers brave men and women whose only crime was to prefer democratic government to U.S.-supported military dictatorships. Worst of all, I watched as the CIA shared its "intelligence" with the leaders of these military regimes, who not unnaturally regarded as an enemy of the state any person named unfavorably in an official CIA report.

As the *New York Times* reported in 1995, "American and Guatemalan officials, who long denied those links, now acknowledge that the CIA gave the Guatemalan military millions of dollars in the 1980s and 1990s, used some of the money as bribes to buy information from high-ranking military intelligence officers, and provided intelligence to the army for its long war against guerrillas, farmers, peasants and other opponents."3

In 1996, President Clinton's Intelligence Oversight Board, assigned to investigate the CIA station's role in Guatemala, found that the Agency had been working at cross-purposes with the Ambassador and that CIA officers had lied to Congress about their activities.

In 1996, I traveled again to Guatemala. A member of the Guatemalan legislature—a man with unassailable democratic credentials—was less impressed than several members of our delegation thought he should be with the Clinton administration's support for democracy and human rights. Yes, he said, Ambassador Marilyn McAfee had carried out an admirable policy designed to advance negotiations toward a peace treaty between the government and the revolutionary movement. He hinted, however, that there were embassy officials who contravened and subverted official policy. To illustrate his point he asked our delegation, "Why is it that the United States is the most stable country in the hemisphere?"

I groaned inwardly, for I knew what was coming. "The United States is the most stable country in the Western hemisphere," he said, "because there is no American embassy in Washington." That joke has been told the length and breadth of Latin America for the past forty years, yet I confess that it still has the power to dismay this former Foreign Service officer.

What possible defense could be made for the CIA's backing of a campaign of torture and terror that ended up killing an estimated 200,000 Guatemalans? There was nothing at stake in Guatemala that could have possibly justified support for such barbarism. Had they been given the opportunity, the American people would certainly have rejected this clandestine war. Covert action, by its very nature, denies citizens their constitutional right to hold public officials accountable for their actions. A better example could not be found for the argument that covert action should be restricted to wartime or when the nation's survival is clearly at stake.

As Marcus Raskin, a former National Security Council official, has written, "A secret bureaucracy, a policy of secrets within secrets, disclosed only to those with 'a need to know,' by its very nature engenders paranoia, institutional ignorance and control by a handful of executives.... The entire project of this community helped to make the government of the United States into an Orwellian nightmare in which 'intelligence' becomes synonymous with massive self-deception and treachery."4

Honduras

In 1962, I was assigned to our embassy in Tegucigalpa, Honduras, as chief of the political section. In my predeparture consultations, I asked the chief analyst of the State Department's Bureau of Intelligence and Research about the classified CIA "Threat Assessment Report" that put the number of communists active in the small country at the astonishingly high level of 5,000. The analyst told me that he, too, had wondered about that statistic, and he asked me to undertake a reality check—because "if that estimate is anywhere near right, Honduras is in deep trouble."

I was fortunate that my Ambassador in Tegucigalpa was John Jova, a gifted career diplomat who insisted on accuracy and precision in reporting. After a year in Honduras, I went to Jova and said, "Mr. Ambassador, I have crisscrossed this country and spoken to every banker, businessperson, politician, priest, union leader, and campesino [agricultural laborer] organizer I could find. I have discreetly cross-checked my information with trustworthy sources and with colleagues in other embassies, and I cannot find 100 communists, let alone 5,000."

I also told Jova that in checking CIA files, I had confirmed that men and women he and I both knew to be responsible, respected professionals had been classified by the CIA as communist sympathizers merely because they had once been student activists and because they now actively opposed the military dictatorship that had come to power by overthrowing one of Latin America's great democratic leaders, President Ramon Villeda Morales.

Both Ambassador Jova and I were well acquainted with the CIA's penchant for exaggerating threats in order to justify operations to overcome fictitious challenges to our national security. Ambassador Jova called in the CIA Station Chief and asked him about the huge disparity between CIA statistics and my research. The CIA representative, an easygoing, professional officer who had inherited the inflated numbers, readily acknowledged that the 5,000 figure was probably exaggerated, but only reluctantly agreed to Ambassador Jova's demand that he document by name every communist in the country. He left the office muttering, "They're not going to like this at Langley"—that is, CIA headquarters.

After six months of evasion and foot-dragging, the Station Chief admitted that the Agency could not come up with more than 200 names. A year later, when the new edition of the Threat Assessment Report came out, the CIA had reduced the number of communists in Honduras from 5,000 to 500, still a gross exaggeration—and chosen, I've always suspected, so that the CIA could claim that the 5,000 figure had been a misprint.

The story has a sequel. The CIA abruptly transferred the candid, cooperative Station Chief and replaced him with a man whose self-importance, high visibility, and arrogance seemed calculated to spite Ambassador Jova for having had the boldness to call into question the CIA's inflated estimates. Within a few months of his arrival, the Station Chief went to see the most influential official in Honduras, the Minister of the Presidency. The CIA official told the minister that he represented the most powerful foreign-affairs agency in the U.S. government, and that if the minister wanted something done, he should work through him, not through the Ambassador.

The Minister of the Presidency, venal and corrupt but an astute politician and no friend of the United States, quickly confided this delicious anecdote to several dozen of his most intimate friends. The Ambassador's prestige and authority were undercut as this breach of discipline within the U.S. embassy became cocktail-party scandal.

El Salvador

In 1979 I was serving as Ambassador to Paraguay. Alarmed by events in Central America, the Carter administration decided that traditional remedies were not working. I was asked to leave Asunción and take over as Chief of Mission of our embassy in El Salvador. Senator Jesse Helms, unsympathetic to my support for democracy and human rights in Paraguay, held up Senate confirmation. By the time I arrived in El Salvador in early 1980, the nation was moving rapidly toward anarchy.

A CIA intelligence assessment titled, "El Salvador: the Potential for Violent Revolution" was published just before I arrived in El Salvador. Its key judgment read, "All of the symptoms associated with impending revolutionary upheaval are present in El Salvador today. Barring major outside intervention, the present social, economic, and political order will collapse or be overwhelmed by the revolutionary left during the next year or two."

Though I did not share the total pessimism of the CIA report, I knew that things were, in fact, coming apart. El Salvador had the worst income inequality and the highest rate of malnutrition in Latin America. The percentage of landless and near-landless was the highest in the world. The military and economic elites that had misgoverned El Salvador throughout the twentieth century believed that death squads and repression were the only answer to a people in rebellion. They were certain that, as in the past, all they had to do was cry "communism" and the United States would rush military aid that would let them crush the revolt their greed had created.

With many years of service in Latin America, I had learned that counterrevolution is never an adequate answer to a people determined to remake their country. The foreign-affairs bureaucracy of the United States has never absorbed that lesson because the Pentagon and CIA have billions to invest in military solutions while the cash-starved Department of State can barely provide its diplomats with the basic tools of the trade.

The crisis in El Salvador and the strong support of Deputy Secretary of State Warren Christopher gave me the rare opportunity to bring U.S. power into play on the side of political, economic, and social change. This would require U.S. backing for profound, far-ranging reforms, respect for human rights, free and fair elections, and, above all, a breaking of the links between the violent right and U.S. policy.

The opportunity to break with the forces of institutionalized violence came quickly and tragically. The most powerful voice in Central America on the side of law and decency was the archbishop of San Salvador, Msgr. Oscar Arnulfo Romero. On March 24, 1980, Msgr. Romero was assassinated while celebrating mass. After discussions with Washington, I spoke out publicly on the murder of the archbishop, placing the blame where it properly belonged, on the violent right wing, with Major Roberto D'Aubuisson as the chief suspect.

For an American Ambassador to condemn right-wing violence in the midst of a left-wing revolution was rare enough to make headlines around the world. Major Salvadoran newspapers, controlled by the people who owned the country, denounced me for daring to speak out on this crime. The *Wall Street Journal* let out an editorial shriek of horror. Yet this public break with the violent right went a long way toward convincing the Salvadoran people that U.S. policy was no longer in the pocket of the military and economic elites and that President Carter had meant what he said when he had called for democratic reform and respect for human rights in El Salvador.

The aim of U.S. policy was to prevent a military takeover by left-wing revolutionary forces. Yet the most immediate threat to the stability of the country came not from the left but from the right-wing death squads, which regularly assassinated moderate politicians, church leaders, and labor representatives. I therefore asked the Station Chief to devote some of his resources to gathering information on the violent right. I made this request in the full knowledge that the CIA had on its payroll agents intimately linked to death squads. My reasonable request met with a point-blank refusal. In the view of the CIA representative, these men were our friends of last resort. He said the CIA's job was to go after the communists, not to spy on our natural allies.

The CIA had spoken clearly. The Agency had its own priorities for El Salvador and did not intend to respond to mine. For me to overlook this challenge would have been to insure the disintegration of a carefully crafted policy. Although I knew that there would be bullying from the CIA and hand-wringing from the State Department, I fired the Station Chief, thereby sending the message that U.S. policy in El Salvador spoke with one voice.

By mid-1980, U.S. policy was beginning to make a difference. I continued to veto all lethal military assistance until the Salvadoran army's human-rights record improved. The government made an honest beginning on agrarian reform, banks were nationalized, and credit began to flow to small farmers and tradesmen. With encouragement from the United States, Napoleon Duarte, the head of the government, responded positively to an offer from Msgr. Romero's successor, Bishop Arturo Rivera y Damas, to bring the government and left-wing political leaders together for negotiations.

The measurable progress away from anarchy and toward a more just society lessened the appeal of the revolutionaries while it infuriated the rich, particularly the coffee barons whose extensive land holdings had been diminished by the first stage of agrarian reform. To these men, El Salvador was a nation to be plundered, not nurtured. They financed one coup attempt after another. As each plot failed, the U.S.-supported government gathered strength and confidence. Yet there was a fatal flaw: the Salvadoran military's addiction to violence.

In the twenty-odd years of my Foreign Service career, I had probably given no more than five or six speeches. In revolutionary El Salvador, however, public statements of U.S. policy became an important tool. I had to make certain that the great majority of Salvadorans, who wanted an end to right-wing oppression but who were frightened of a violent takeover by left-wing revolutionaries, understood that U.S. policy had dramatically changed.

It was time for a major statement reaffirming U.S. commitment to economic reform and denouncing death-squad violence. I circulated a draft of the speech to members of the country team. The acting CIA Station Chief came to my office the morning the statement was to be delivered and told me that a death squad had assassinated a left-wing political leader, Juan Chacon. This report saddened me; like so many young Salvadorans who had believed in democracy, Chacon had joined the revolution only because he had become convinced that all peaceful options for change had been exhausted.

The acting Station Chief said that CIA headquarters strongly recommended that my speech condemn the killing of Chacon. Although I welcomed this first signal that the CIA was willing to cooperate with the new policy, I was still cautious. Any attack on official violence had to rest on a solid factual base. Any confusion or misstatement would hold both the policy and me up to ridicule.

I thanked the acting Station Chief for the recommendation but told him that I did not want to take any risks. He responded that Langley was most insistent that this represented an opportunity to drive home the policy with new facts. I asked that he reconfirm the report. He returned in an hour, assuring me that the death of Chacon had been confirmed and reconfirmed. I can still recall the collective gasp that went through the audience when I introduced Chacon's name as the most recent victim of military death squads.

I had fallen into a CIA trap. Chacon was not dead. I had committed the fundamental error of considering the CIA station as part of the embassy team committed to presidential policy. I had underestimated the CIA's hostility to any course of action in Central America that did not include the Agency as a major player. Nor had I reckoned on the CIA's vengeance against any Ambassador with the mettle to exercise his authority as the President's representative and expel a Station Chief who refused to respond to policy priorities.

To make a bureaucratic battle of this betrayal would have proved worse than useless. Few high-ranking State Department officials had the stomach to take on the CIA, and they would have correctly judged me as having been too trusting. I had no alternative but to turn my attention to repairing the considerable damage and getting on with the job.

By October 1980, the leftist threat, while still dangerous, had measurably receded. In stark contrast to its earlier predictions, a CIA analysis concluded that there was now sufficient governmental stability that any imminent threat of a leftist takeover could be ruled out. The Carter policy, with its emphasis on reform, human rights, and political solutions, had begun to take hold.

With the election of Ronald Reagan, this approach, so full of promise, was jettisoned for the unattainable goal of military victory. A quick, decisive crushing of the Salvadoran insurgents became the battle cry. "Mr. President, this is one you can win," said Secretary of State Alexander Haig. It took ten years, 75,000 murdered civilians, and a million Salvadoran emigrants to pry U.S. policy loose from Haig's delusion.

It should have been the job of CIA analysts to blunt such uninformed zeal with objective reports that warned U.S. policymakers against sinking deeper into this Central American morass. In this basic responsibility, the CIA failed. As had happened too often in the past, the CIA put intelligence at the service of policy and provided optimistic, misleading evaluations that served as justifications for increasing involvement in El Salvador. The one indisputable result of Reagan's policy was the emergence of the strongest, most resourceful guerrilla movement ever seen in Latin America.

Many Foreign Service officers did what they could to bring some sense of balance and proportion to our presence in Central America. We failed because at the center of our policy was a doctrine of national security that relied on a strategy of counterinsurgency. The Pentagon and the CIA taught the Central American armies not only to attack guerrillas but also to suppress those who insisted on using their rights of free speech and free assembly to dissent from official policy. We also failed because the CIA agents who wrote murder manuals and trained death squads in several countries were accountable to no one except their bosses in Langley.

When the CIA was born, Secretary of State Dean Acheson expressed "the gravest foreboding" about the Agency, warning that once in being, "neither the President nor anyone else would be in a position to know what it was doing or control it." Many Ambassadors throughout the world have had reason to echo Acheson's words.

Haiti

When the CIA insists that it never undertakes covert action without presidential authority, recall the recent case of Haiti, where CIA actions clearly went *against* presidential policy. President Clinton had made it clear from the outset of his administration that he would not tolerate the continuation in power of the military leaders who in 1991 overthrew Haiti's first elected President, Jean-Bertrand Aristide. Clinton said, "I want to emphasize how important it is to me personally to restore the democratic government to Haiti and how important it is to the United States that we return President Aristide to power." No empty phrasing here. No ambiguities that could justify competing interpretations or backsliding among Clinton's foreign-policy team.

Yet the CIA was not willing to abandon its longtime clients, the military generals of Haiti. The National Intelligence Officer for Latin America, Brian Latell, performed a slash-and-burn operation on Aristide in the form of a psychological profile provided to Senator Jesse Helms. According to Latell, Aristide had undergone psychiatric treatment in a Canadian hospital. This claim and others were later proved to be false.

A *Miami Herald* editorial on November 7, 1993, stated that the CIA was circulating "scurrilous, unsubstantiated drivel about Haiti's President Jean Bertrand Aristide." The same editorial concluded that "the CIA has been, and is, impeding the Clinton administration's efforts to restore Aristide to his presidency."

The CIA director, James Woolsey, then revealed under congressional questioning that the Agency had kept on the payroll key members of the Haitian military high command after they had presided over the overthrow of Aristide. This revelation made more comprehensible the unyielding defiance of Haitian army leaders to the return of the elected President. It was obvious to the Haitian generals that if the CIA continued to pay them after they had ousted the constitutional government, then the United States government secretly approved of their action, despite its public statements. The generals knew they had defenders in the American foreign-policy community, and that no serious move against their power grab was to be expected.

Perhaps most indefensibly, the CIA in Port-au-Prince recruited Emmanuel Constant, the leader of FRAPH—the paramilitary Revolutionary Front for the Advancement and Progress of Haiti—which spread terror throughout the country by persecuting and murdering Aristide's supporters in the weeks prior to his scheduled return. CIA sponsorship of this terrorist again confirmed to the Haitian military that the administration would not insist on Aristide's return.

After months of these contradictory signals, President Clinton had no alternative but to send American troops to oust the Haitian military junta and restore Aristide to office. Clinton was driven to these extreme measures at least in part because the CIA worked against the presidential policy of achieving and carrying out a negotiated solution.

CIA activities in foreign countries are usually described as falling into two categories: clandestine collection of information and covert action. The latter requires a presidential authorization known as a "finding." In my experience, there exists a third category, a hybrid that parades as information-gathering but in reality is a form of covert action. The CIA contends that it has no choice but to recruit uniformed criminals such as General Manuel Noriega of Panama and political assassins such as Emmanuel Constant in order to gather intelligence. This claim is false and self-serving. These tropical gangsters enjoyed profitable contractual arrangements with the CIA not because they were particularly good sources of information but because they served as paid agents of influence who promoted actions or policies favored by the Agency.

Proposals for covert action may come from the CIA station in a given country, from CIA headquarters, or, occasionally and usually disastrously, from the White House. Examples of the CIA going directly against the White House, such as Haiti, are rare. More typically, the CIA is less a "rogue elephant," as Senator Frank Church once called it, than a unique bureaucracy operating in secret with seemingly limitless funds and no pesky Congress or public able to call it to account. The existence of such an agency acts as a perpetual temptation for Presidents and their chief aides to look for a shortcut rather than undertake the long, arduous task of building a consensus within the executive branch and the Congress for a set of policies designed to meet a foreign challenge. The CIA offers the illusion of a cheap, quick, and easy fix.

CUBA

This approach to foreign affairs can seriously backfire. In Cuba, the United States provided Fidel Castro the gift of the monumental fiasco of the Bay of Pigs, eight failed assassination plots, and a succession of conspiracies designed to sabotage Cuba's economic base. As every CIA agent recruited inside Cuba became a double agent reporting to Castro, the CIA lurched from one embarrassing failure to the next, with the result of entrenching Castro in power.

It takes a satirist of the stature of Joan Didion to capture the zany, delusional quality of this and so many other CIA operations. In her book *Miami,* Didion wrote:

> According to reports published in 1975 and 1976 and prompted by hearings before the Church Committee, the Senate Select Committee to Study Governmental Operations with Respect to Intelligence Operations, the CIA's JM/WAVE station on the University of Miami campus was by 1962 the largest CIA installation, outside Langley, in the world, and one of the largest employers in the state of Florida. There were said to have been at JM/WAVE headquarters between 300 and 400 case officers from the CIA's clandestine services branch. Each case officer was said to have run between four and ten Cuban "principal agents," who were referred to in code as "amots." Each principal agent was said to have run in turn between ten and thirty "regular agents," again mainly exiles.
>
> The arithmetic here is impressive. Even the minimum figures, 300 case officers each running 4 principal agents who in turn ran 10 regular agents, yield 12,000 regular agents. The maximum figures yield 120,000 regular agents each of whom might be presumed to have contacts of his own. There were, all operating under the JM/WAVE umbrella, flotillas of small boats. There were mother ships, disguised as merchant vessels, what an unidentified CIA source described to the *Herald* as "the third largest navy in the western hemisphere." There was the CIA's Miami airline, Southern Air Transport, acquired in 1960 and subsequently financed through its holding company, Actus Technology Inc., and through another CIA holding company, the Pacific Corporation, with more than $16.7 million in loans from the CIA's Air America and an additional $6.6 million from the Manufacturers Hanover Trust Company. There were hundreds of pieces of Miami real estate, residential bungalows maintained as safe houses, waterfront properties maintained as safe harbors. There were, besides the phantom 'Zenith Technological Services' that was JM/WAVE headquarters itself, fifty-four other front businesses, providing employment and cover and var-

ious services required by JM/WAVE operations. There were CIA boat shops. There were CIA gun shops. There were CIA travel agencies and there were CIA real-estate agencies and there were CIA detective agencies.[5]

It now becomes clear how a CIA operation designed to overthrow Cuba's new revolutionary government accomplished the rare feat of serving the opposing interests of both sides. JM/WAVE hired and paid thousands of Cuban exiles and pumped tens of millions of dollars into the south Florida economy, thereby bringing affluence to many members of the newly arrived Cuban-American community. The CIA effort also served the interests of Fidel Castro, for revolutions gather force and legitimacy from their enemies. The United States had long treated Cuba as a virtual colony. Freedom-loving Cubans would never have sacrificed their liberties to the Cuban government except under the threat of a foreign invader. The Bay of Pigs and JM/WAVE solidified Castro's grip on power.

It was once said of Philip II, the king who sent the Spanish Armada to its destruction, that no experience of the failure of his policy could shake his belief in its essential excellence. A fitting epitaph for the CIA's inane adventurism against Cuba.

Nicaragua

The 1979 Sandinista victory over the tyranny of Nicaragua's ruler, General Anastasio Somoza, began the mildest revolution Latin America had ever seen. Unlike the Cuban revolution, the Sandinistas did not resort to drumhead trials and summary executions, did not repudiate the enormous debts piled up by the Somoza kleptocracy, and did not quit the Organization of American States. They clearly signaled revolutionary Nicaragua's intent to live by the rules of the inter-American community.

The overwhelming majority of Sandinistas understood that the future of Nicaragua depended on establishing good working relationships with the United States. The revolutionary government committed its share of follies, blunders, and excesses, as is inevitable when young, idealistic men and women take over the direction of a country by force of arms. Yet time and again, the Sandinistas offered to negotiate with the United States, assuring Washington that they would never permit foreign bases on Nicaraguan soil and giving a commitment to the U.S. Ambassador that they would cease sending arms and material to the Salvadoran revolutionaries.

With Director of Central Intelligence William Casey in the lead, the Reagan administration brushed aside Sandinista peace overtures and in 1981 began formulating plans to overthrow the Nicaraguan government. The CIA created, trained, and paid a counterrevolutionary army, composed largely of former members of the hated Somoza National Guard, and forced the government of Honduras to permit the Contras to use its territory. Over the next five years, the Reagan administration attempted to conceal its policies and actions from the public. After it was revealed that the CIA had mined the harbors in Nicaragua in 1984 and prepared a training manual instructing the Contras how to assassinate judges and other troublesome officials, a frustrated Congress cut off all funding to the Nicaraguan Contras in order to put an end to this proxy war.

Unwilling to permit the Congress to kill his dream of ousting the Sandinistas, President Reagan authorized the CIA to sell arms to Iran and to use the profits to support the Contras. This presidential action appeared to violate Congress's exclusive right to control the purse, as stated in the Constitution. According to the independent counsel on Iran-Contra, Lawrence Walsh, this action brought President Reagan "within range of impeachment." The timely death of William Casey removed the key witness to Reagan's direct involvement, and a constitutional crisis was averted.

Even so, the number and rank of officials brought low by Iran-Contra was impressive: Secretary of Defense Caspar Weinberger was indicted; National Security Adviser Robert MacFarlane pleaded guilty and a received a suspended sentence; National Security Adviser John Poindexter and his aide, Oliver North, were convicted on multiple felony accounts; a plea bargain reduced felony charges to misdemeanors for Assistant Secretary of State Elliott Abrams and Alan Fiers, a CIA officer; Clair George, a CIA official, was convicted of a felony; and another CIA officer, Dewey Clarridge, was indicted. In what Walsh called "the last card in the cover-up," President George Bush, the former director of the CIA, issued what might be described as a clandestine presidential pardon for the chief offenders, signed on Christmas Eve 1992, with no press or photographers allowed.

Few regions of the world have ever suffered such concentrated death and destruction as did Central America in the 1980s, as a result of proxy armies fostered by the United States: more than 125,000 civilians killed; tens of millions of dollars of property and productive assets destroyed; the environment devastated; 2 million refugees, desperate to escape the violence, fleeing to our borders and entering the United States as illegal immigrants.

Throughout the decade, Washington steadfastly beat back opportunities for honorable and constructive negotiations designed to satisfy all U.S. security requirements. It was simply dishonest for the Reagan foreign-policy team to charge that the Sandinistas had "betrayed their revolution" when the administration had immediately put the CIA to work to make certain that the Nicaraguan revolution would fail.

Central America began assuming responsibility for its own security with the 1988 signing of the Esquipulas II peace treaty. With President Oscar Arias of Costa Rica in the lead, five presidents turned thumbs down on a Reagan administration proposal designed to achieve at the negotiating table what it had not been able to gain on the battlefield, and instead agreed on a pact of their own design. They decided simultaneously to end all outside help to insurgent forces, to put a stop to all use of their territories for attacks against neighboring countries, and to seek cease-fires, amnesties, political dialogue, and democratization.

Esquipulas marked the beginning of a new Central America—one no longer subservient to the cycles of U.S. intervention and neglect. Over the course of years, the peace process has created new regional mechanisms for preventive diplomacy and mutual support. U.S. policymakers should welcome these signs of maturity and responsibility, and should put their influence to work by unambiguously backing civilian authority and giving support to Central America's own security initiatives. As Adam Isacson wrote in *Altered States,* his fine book on security and demilitarization in Central America, "New directions in United States policy can decisively shift Central America's civil-military balance toward civilians. While it is possible that the lasting achievement of both security and civilian control can occur without open U.S. support, the obstacles would be much greater and the chance of success much lower. We urge the United States government to break with its history by building an alliance with Central America's democratizing sectors and supporting regional demilitarization."[6]

In Central America and the Caribbean, espionage and clandestine collection of information are a waste of both time and money. To supply our country's basic informational needs, the first requirement is an embassy in each country headed by a capable Ambassador who enjoys the confidence of the President and other high Washington officials. These diplomatic offices should be staffed by Foreign Service officers who have a profound understanding of the host country's people, language, and problems.

Second, there should be a central intelligence agency located in Washington, staffed by a corps of knowledgeable analysts but with no policy or operational responsibilities. As befits an open, democratic society, there should be a constant flow of communication between the government and a network of university research centers whose independent analyses can spur the thinking of policymakers who would otherwise spend too much time inhaling one another's stale ideas.

It is one of the primary tasks of an embassy to report accurately and in depth on issues of consequence to Washington. This is not as difficult as it may appear, particularly in Central America and the Caribbean. Government officials, business people, politicians, and religious and social leaders all want to understand the policy of the United States toward their country. It is the give-and-take of these discussions that enables the Foreign Service officer to assess the capabilities and intentions of the country's leadership.

By presidential order dating from early Cold War days, embassies must provide cover for CIA operatives, although that cover is usually deliberately transparent. But instead of concentrating on important threats to our national security, the swollen bureaucracy of the CIA has imposed huge numbers of its operatives on U.S. embassies in countries with open societies. In many overseas posts, CIA officers outnumber legitimate diplomats.

The one commodity never in short supply in Central America and the Caribbean is information. What the President and minister of defense discuss in the morning, the market women gossip about in the afternoon. Why pay for something that is available at no cost?

Information acquired free through nonclandestine sources has another huge benefit: It carries no risk of betrayal. Yet in the competition for the foreign-policy dollar, overt diplomacy has come up the loser and clandestine collection the clear winner. Budget cuts have forced the State Department to close more than thirty embassies and consulates and shut down USIA libraries around the world, while the budgets of intelligence agencies sail through Congress. The Israeli statesman Abba Eban was surely correct when he recommended "a less deferential attitude to covert diplomacy in all free societies. These are more colorful than regular statecraft and far more productive of best-selling novels and investigative journalism, but they rarely affect the larger currents of history. More work for foreign offices and less for intelligence agencies would give mankind a measure of relief."[7]

Central America and the Caribbean is the ideal region to put Eban's recommendation into practice. Withdraw all CIA staffs from these small countries, and let career diplomats be charged with fulfilling Washington's intelligence requirements. Should Foreign Service officers prove capable of meeting all intelligence needs, gradually extend this beneficial practice to all democratic countries in Latin America. This could be accomplished through pacts of reciprocal restraint in which signatories would agree not to spy or engage in covert action against the other. Central America and the Caribbean could serve as a test case, letting us gauge the need to station CIA operatives in democratic countries—beyond, perhaps, minimal needs for liaison with other intelligence services.

How a country is organized to conduct its foreign policy determines, to a great extent, the nature of that policy. Over the past four decades, we have overfunded spying and underfunded diplomacy. As a result, United States policy toward Central America and the Caribbean has too frequently departed from the path of judicious application of diplomatic influence in pursuit of legitimate objectives. We have trampled on sovereignty, judicial equality, and territorial integrity, principles that served for centuries as norms of conduct among independent states. It is time to reassert these fundamental principles and give them the preeminence that they deserve by setting aside covert means—at least until legitimate, responsible diplomacy has been tried and found wanting.

Notes

1. Daniel Patrick Moynihan, *On the Law of Nations* (Cambridge, MA: Harvard University Press, 1990), p. 134.

2. *New York Times,* February 26, March 1, and March 7, 1999.

3. *New York Times,* April 2, 1995. See also February 26, 1999.

4. Marcus Raskin, *Abolishing the War System* (Northampton, MA: Aletheia Press, 1992), p. 92.

5. Joan Didion, *Miami* (New York: Vintage Books, 1998), pp. 90–91.

6. Adam Isacson, *Altered States: Security and Demilitarization in Central America* (a joint project of the Center for International Policy, Washington, DC, and the Arias Foundation for Peace and Human Progress, San José, Costa Rica, 1997), p. 200.

7. Abba Eban, quoted in the *New York Times,* December 30, 1986.

Robert V. Keeley

4   CIA–Foreign Service Relations

Why do we need spooks wearing cowboy boots when we've got perfectly capable cookiepushers in striped pants?[1]

It is said that spying is the second-oldest profession. In most cultures its moral standing is equal to that of the oldest profession: prostitution. Diplomacy might rate as the third-oldest profession. The defining difference between espionage and diplomacy is that the former is a tactic used in political rivalry or actual military warfare, while the latter is a tactic used in conflict resolution or actual peacemaking.

Both espionage and diplomacy have always engaged in gathering information—what today is labeled "intelligence"—spies obtaining it clandestinely and diplomats acquiring it openly. The immense growth of the American spying apparatus since World War II is generally attributed to the debacle of the Japanese sneak attack on Pearl Harbor in December 1941, a devastating loss that many believe would have been avoidable had we benefited from adequate intelligence. Avoiding another such catastrophe justified the creation of the peacetime CIA (in part a successor to the wartime OSS) and the consequent establishment of small to very large contingents of American intelligence officers in most of our embassies and consulates throughout the world.

This melding of spies and diplomats in our official establishments abroad has created a host of problems, mostly to the detriment of our professional diplomats, the officers of the Foreign Service (known as FSOs). The justification for placing the CIA's "stations," as they are called, in our diplomatic and consular establishments is basically twofold. The most important reason is to protect U.S. spies who are engaging in illegal activities in the country. If caught, they could face long prison terms or even execution. As protection, they are given diplomatic "cover"—that is, they are accredited to the host government as regular diplomats, just like FSOs, and therefore enjoy diplomatic immunity. The only recourse a foreign government has for dealing with a U.S. spy on embassy or consulate rolls is to declare him persona non grata (officially unwelcome) and demand his removal from the country, a requirement that all governments must honor, according to diplomatic protocol.

The second justification for housing our spies in our diplomatic establishments is that it provides security for their files and communications, as diplomatic premises are supposed to be inviolate from intrusion.

All countries, of course, use similar arrangements to protect their spies and their means of communication with their headquarters. But we are here concerned with the American experience.

The CIA also places some of its personnel abroad in positions outside embassies and consulates, in positions that can legitimize their presence in a foreign country. They may be businessmen, bankers, scholars, employees of international or nonprofit charitable organizations, or associated with any other nongovernmental entity that can provide what is known as "nonofficial cover." Some spies are given "deep cover," which simply means that the professional or other disguise they use is so good that their true profession as U.S. spies is unlikely to be suspected.

The practice of clothing clandestine intelligence officers in diplomatic cover is so well known that foreign governments make an effort to identify which American diplomats in their country are actually spies. They often succeed, and they often make mistakes. In countries where the CIA has established a "liaison" relationship with the host government's intelligence service, some American CIA personnel are openly identified to the host service, an obvious necessity if the two services are to exchange information deemed of mutual benefit or to cooperate in other endeavors—to work against terrorists, for instance. In a significant number of countries, the host intelligence service is actually a creature of the CIA, organized, trained, and even funded, at least in part, by the Agency. In these cases, of course, the relationship is likely to be close and extensive.

It should not surprise anyone that such relationships between spying agencies exist even with countries that are neither American allies nor especially friendly to the United States. Since the host government realizes that it is going to be spied upon by the CIA and can do nothing to prevent that, a liaison relationship has the appeal of at least providing a channel for keeping some tabs on CIA activities and for protesting if they get out of hand.

The most obvious negative consequence of mixing spies with diplomats is that it automatically places the latter under suspicion of being spies. Foreign officials, ordinary citizens, and all other potentially useful contacts or interlocutors for a diplomat are likely to be wary of dealing with someone they think may be a spy. So in providing diplomatic cover to spies to protect them from prosecution, we are at the same time painting our diplomats with a presumed espionage patina that can seriously impede their ability to function in their true diplomatic role.

Another negative consequence is that the proliferation of spies abroad creates duplication and a waste of resources. Diplomats assigned abroad as political, economic, and politico-military reporting officers provide most of the foreign open-source information that supports the intelligence estimates provided to Washington policymakers. In recent years this reporting function has been extended to cover such topics as terrorism, narcotics trafficking, crime, the environment, and nuclear proliferation. At the same time, hordes of American intelligence officers are supplying information on these same areas, most of it obtained clandestinely.

Many FSOs could provide examples of the most extreme form of overlap by citing personal cases in which they developed very useful foreign sources of human intelligence (HUMINT, in CIA jargon), only to find a source drying up because he had been recruited by a CIA case officer and put on the CIA's payroll. Uncle Sam was now paying for what we had been receiving free.

The CIA argues that its paid, clandestine sources are more valuable once they are on the CIA payroll because they are now "controlled"—under CIA control, that is. This is a false concept, as tautological as I have expressed it, and is more serious than the waste mentioned above. Once on the payroll, the controlled source has a strong motivation to tell his controller what he thinks the latter wants to hear, because if he doesn't he may not stay on the payroll for long. The result is that the CIA's clandestinely obtained information most often reinforces the worldview of the CIA—and let no one claim there is no such thing as a CIA worldview.

An open source, such as an FSO's contact, may have his own ax to grind and may well provide misleading information, but that can be evaluated objectively, just like all other information. On the other hand, the CIA case officer who has recruited a "controlled source" has an incentive to insist that his recruit's information is as good as gold.

The most pernicious effect of this competition for sources is that for a number of reasons—the CIA culture, the cult of secrecy, and the general tendency of naive politicians in both the executive branch and Congress to be thrilled when they are allowed access to allegedly important "secrets"—a much higher value is placed on clandestinely obtained intelligence than on what is received free from open, unpaid sources. After all, don't we all place much greater value on what we have to pay for than on what we receive free? This is a fact of our socioeconomic culture: We value products that are vastly overpriced simply because they are expensive, whereas we don't hesitate to dump a free handout in the nearest trash can.

I can't stress enough the mistakes we have made by romanticizing expensively obtained information over what is in the public domain. The problem today in the field of intelligence is that there is such a surfeit of information freely obtainable throughout the world. Indeed, it is ridiculous in most cases to try to buy intelligence allegedly obtainable in no other way, when our most urgent task is to sift through the avalanche of freely circulating material to find the nuggets that help us understand what is really going on.

This situation makes the CIA very vulnerable to being tricked into recruiting what turn out to be "double agents," who are working for both our intelligence service and the enemy's. Egregious examples are described elsewhere in this book. (See Chapter 2.) The bottom line is that we have been paying for false information, and, even worse, false information designed to deceive us by agents of an adversary intelligence service. There is a long record of deception, lying, and fraud in the history of regular diplomacy, but it has been mitigated by the skepticism normally accorded information freely supplied by foreign sources. That skepticism tends to disappear when we are disbursing taxpayer dollars for "intelligence" of dubious origin and motivation.

CIA officers tend to recruit sources who are in positions of power and influence, giving them access to allegedly valuable information. These people are often members and servants of the ruling regime. They are also the normal contacts of regular diplomats. Aside from the overlap and redundancy problem, this anomaly creates a situation in which the CIA neglects to report on the opposition while hobnobbing with the ins. Thus it is left to the diplomats to contact, deal with, and report on the opposition, the outs. The latter may be considered subversive or even revolutionary (and thus illegal) by the ruling regime. In many cases, the regime has been installed or at least encouraged to take over by its friends in the CIA. We are now operating in a topsy-turvy, *Alice in Wonderland* world in which our clandestine operatives are working with the often oppressive ruling clique, and the regular diplomats are working with the often democratic opposition forces. The correct division of labor would have the diplomats dealing with the government in power, to which they are openly accredited, while the CIA collectors would report on the opposition, which often can be reached only by clandestine means.

In cases where a country is ruled by a CIA-supported regime, the Agency has a vested interest in making that regime look good, even when it is oppressive and headed for eventual overthrow. This is a consequence of placing covert action and clandestine collection in the same agency. Too much CIA reporting is tainted by its bias toward governments—of whatever stripe—that are guaranteed to support, or at least tolerate, the Agency's operations and objectives.

In a country where the CIA has embraced, or even installed, the ruler or a ruling clique, that leader can hardly be blamed for believing that the real power in Washington is wielded by the CIA. This is an impression buttressed by the Agency's ability to provide resources outside normal channels, whether cash, arms, political support, public-relations advice, or information about the domestic opposition. Simply put, such a regime concludes that it's better to channel its dealings with the United States through the CIA rather than through the Ambassador and his embassy because the CIA has more influence over policy in Washington. American Presidents, if they are aware of such situations, have rarely stepped in to back their appointed personal representatives against an aggressive CIA challenge. This not only distorts relationships between governments but can cause a great deal of dangerous confusion on all sides, especially when an embassy and its CIA station are pursuing different, even contradictory, policies.

In the past, efforts have been made to deal with this problem, notably by President Kennedy's letter to all Ambassadors placing them in complete charge of all official U.S. government activities and personnel in their country of accreditation (with the exception of military personnel serving under an area commander who reports directly to the Pentagon).[2] President Carter tried to reinforce Ambassadors' control over CIA activities by requiring the CIA Station Chief to inform the Chief of Mission about any CIA sources whom the Ambassador might be dealing with in his or her normal course of business with the host government. In my experience, this breach in the legally authorized stone wall behind which the CIA is allowed to protect its "sources and methods" from disclosure to anyone outside the Agency has been handled with exceptional flexibility by many Station Chiefs. Typically, they wait until the Ambassador has asked the right question—"Is the crown prince on your payroll?"—before providing the required information. Volunteering such a disclosure is a rarity.

Ambassadors are still denied any access to the Agency's communications between stations and headquarters that are deemed to be "operational," and even reports that are purely "informational"—"intelligence," that is—are protected under the "sources and methods" stricture. As for covert actions, these are shared with outsiders, if they are shared at all, only on a "need to know" basis, with the decision left entirely in the hands of CIA personnel. It's no wonder that most Ambassadors have rarely been able to exercise serious oversight over CIA activities in their host countries.

Several Station Chiefs I have worked with have confided to me that their very highest priority abroad was to engineer defections of senior Soviet officials, especially officers of the KGB and GRU, and similar officials of other communist governments. This came ahead of any intelligence-collection function, and a successful recruitment could make any officer's career bloom. Much lower priority was accorded to reporting on the country where the station was located.

One reason for the extreme focus on recruitment of enemy agents and defections of enemy officers, especially intelligence personnel, was that the CIA desperately wanted to know whether there were "moles" in the CIA or our other intelligence agencies: American personnel secretly working for the enemy. The Soviets and their allies, particularly the East Germans, did the same, giving very high priority to recruiting CIA officers to uncover U.S.-government-recruited moles in their intelligence agencies. Vast resources on both sides were devoted to penetrating the rival services, to placing moles there, and to engineering defections, all for the purpose of uncovering penetrations of one's own service and the placement of moles within it.

Finally, it has been most distressing to the Foreign Service in recent decades to observe the severe cuts made in the resources that our government has devoted to regular diplomacy and its adjunct activities, such as foreign aid, cultural exchanges, and international organizations. These appropriations have been cut by about 50 percent, and diplomatic staffing levels have been affected most of all. This has come at a time of turbulence in international affairs; instability in many countries and entire areas, such as Africa and Asia; and ever-growing demands on the Foreign Service to report on such areas of interest as narcotics, crime, and nuclear proliferation. In the same period, there has been a natural growth in the number of countries we have to deal with; today the number is 174.[3] The Foreign Service has been shrinking at a time when it is being called upon for more production than ever before.[4]

Today we are relying more on the CIA for information at a time when clandestine sources are in most cases needed less than in the past and when the availability of open-source information has grown exponentially. There is danger in our relying too much on spies and too little on diplomats for the foreign information and relationships needed today to protect our national security. Both the executive branch and Congress are to blame for authorizing this shift in priorities.

What follow are real-life illustrations of some of the problems discussed above. They are not intended to present a uniformly negative picture of CIA activities overseas, but they add anecdotal information to what I believe is a clear case for reform of CIA–Foreign Service relations.

My first story illustrates how the greater resources of the CIA in the field result in the Agency securing sources that then become unavailable to the Foreign Service and out of the orbit of open diplomacy. In an African country, shortly after it attained its independence, I befriended a young member of the cabinet who had been a leading figure in the independence struggle. He was an unusually open person, inexperienced in politics, and he saw nothing wrong with being informative with an American diplomat about his country's new situation. My wife and I attended his wedding in his home district, and my wife was helpful to the bride in settling into her new role.

One day, the cabinet minister confided to me that he was having trouble meeting his mortgage payments. His salary was small, but because of his high position in the new government he was being besieged by relatives and constituents for handouts, government jobs, places in the university, loans to set up businesses, and other difficult-to-refuse financial demands. His mother and some other relatives had even moved into his house, and his food and drink bills were bankrupting him. All this distressed his young wife no end, and the final straw for her—she confided to my wife—came when she slipped away to their bedroom one night in an exhausted state only to find that three of her husband's cousins were already asleep in the marital bed. Her husband naively asked me whether the embassy could help him with his mortgage payments.

I explained that we didn't have funds for such purposes, but, without telling him, I informed our CIA Station Chief of this situation, knowing that he would follow up on it immediately. As I expected, the minister was soon on the CIA payroll and we had a clandestine pipeline into the host government's cabinet. As I'd also expected, the minister dried up as a primary source for me, though we remained friends and continued to exchange views. Our talks were an education for me in the problems of being a politician in a democratic country in the new Africa, and this was highly useful. But I was no longer the recipient of the inside dope on cabinet deliberations and decisions—information that we were now purchasing.

At a newly established embassy in Africa, where I was assigned, we had no Marine guards to stand watches in our chancery to protect our classified files and codes. Our Ambassador insisted that all of his officers take turns sleeping in the chancery, one night a week per officer, as substitute guards. One of our chores while standing these watches was to inspect all offices to make sure no classified materials had been left exposed instead of locked in safes. This scouring of the chancery included checking the paper trash thrown into office wastebaskets.

On one of my night duties I found a most unfortunate discard in the CIA officer's wastebasket. It was a copy of an uncoded draft of an "operational" message to CIA headquarters, suggesting that a new junior Foreign Service officer recently arrived at the embassy might be recruited to become an undercover CIA officer in his spare time, given that he had served briefly in the CIA in Washington before joining the Foreign Service.

Needless to say, I was shocked. All I did was lock up the wayward message in my safe; in the morning I turned it over to my CIA colleague in the embassy without comment. His secretary—the culprit in this serious breach of security—was mortified. I don't know to this day how many people, if any, got into trouble over this episode, but ever afterward it made me wonder how many of my FSO colleagues might also be "double agents" working undercover for the CIA.

In yet another African country, this one ruled by one of the most brutal military dictators in recent African history, we had "enjoyed" a direct relationship with the general before he came to power, as he had been a principal "asset" of the CIA for some time. Although the Agency had little if anything to do with his coming to power in a bloodless coup, the United States, along with Britain and other Western powers, welcomed his advent in replacing an increasingly left-leaning and uncooperative elected President. The CIA, of course, welcomed the idea of one of its assets occupying the throne of power.

We were swiftly disillusioned, as the mercurial dictator began veering off in hostile and vicious directions. Most of his brutalities were inflicted on his own people, but his overall posture was also taking on a distinctly anti-Western coloration. He soon cut off his contact with the CIA and began denouncing the Agency as an enemy organization trying to subvert his country and, of course, his own rule. He lumped the CIA in with imperialists, Zionists, and other alleged enemies, and began denouncing missionaries and Peace Corps volunteers as Agency operatives. The CIA officer who had previously served as his contact had to be transferred away for his own safety. What had for some years been a useful liaison with a powerful figure of the previous regime—the head of the army—became a nightmare when that person turned out to be a murderous tyrant.

The low point for me personally in this country occurred a couple of years later, when I was serving as chargé d'affaires following the recall of our Ambassador after a serious deterioration in relations. In an attempt to humiliate my government, our embassy, and me personally, the general exposed on television a room in a house from which the CIA had been tapping the telephones of the Soviet and Chinese embassies. He made me witness this scene with the Soviet Ambassador and the Chinese chargé. He identified for the TV audience the author of this bugging scheme as the American CIA case officer who had been his own contact with the Agency. This was a connection that he, of course, suppressed, as he did the fact that he had personally authorized this phone-tapping operation in his previous incarnation as the army chief, at a time when he thought his enemies were communists rather than imperialists. The case illustrated for me the dangers of recruiting leaders whose values are antithetical to ours, despite the convenience of having a useful source or ally in the country.

The Congo

In two other African cases I can identify the country because my involvement was from the Washington end, as a desk officer. The Congo is nothing less than a CIA horror story. Throughout his long and infamous career, Mobutu Sese Seko maintained close ties with the CIA, and most of his important dealings with the U.S. government were handled through CIA channels. When we had some bad news to deliver to him, however, such as turning down his latest request for some expensive and dangerous military toys, it was always the Ambassador who had to deliver the unwelcome message. In at least three such instances, Mobutu's reaction was to demand the withdrawal of the American Ambassador. In every case, Washington complied and sent him a new Ambassador instead of acting as it should have: downgrading our representation to chargé level, shutting down the cozy CIA-Mobutu relationship, and insisting that all government-to-government exchanges be placed in diplomatic channels.

For more than thirty years, Mobutu ran a kleptocracy so effective, refined, extensive, and brutal that it has come to define that kind of regime. His legacy has been the transformation of a potentially prosperous country into a basket case that will probably require as many years to recover from his rule as it took him to devastate it—if it ever does recover. No doubt Mobutu was a useful and cooperative tool ("ally") in our Cold War competition with the Soviets for influence in Africa, but the cost was excessive—stupendous, in fact. And these were not unintended consequences of a CIA operation that became visible only after the fact, as one could argue about Afghanistan, for example. Everyone knew what Mobutu was doing to his country, from the very beginning to the very end. The CIA and the U.S. government in general evidently didn't care. The lesson here is: Don't put the CIA in charge of your relations with any country.

Ethiopia

Ethiopia nicely illustrates the point about our excessive reliance on relations with the ins when it should be the CIA's job to keep close tabs on the outs, lest they come to power and leave us lacking understanding and relationships.

I took charge of the Ethiopian desk at the State Department in Washington about a week before the coup d'état in January 1974 that ended the long rule of Emperor Haile Selassie and his entire imperial regime. All of Washington was shocked by the overthrow of the Emperor, and it was imperative to find out as soon as possible who had done the deed, why, what they stood for, and where they intended to take the country, which was vital to our interests in the region (one of our few military bases in the area was at Asmara) and to our interests in Africa in general.

The pressure for an immediate assessment of this sudden turn of events was compounded by the fact that both President Richard Nixon and his new Secretary of State, Henry Kissinger, considered Haile Selassie a personal friend. (One of Kissinger's major failings was his personalization of the country's foreign relations, as if they were his own foreign relations.)

I was assigned to draft a "contingency paper" addressing the policy issues and implications now facing the U.S. government, to be read by senior officials in the State Department, the Pentagon, the CIA, and the White House. My difficulty in drafting the paper was that we knew very little about the character and probable intentions of the new leaders, a military junta of colonels and other middle-grade officers, with an apparently figurehead general installed at the top.

Here I could gain little of use from the CIA, as it had been totally in bed with the major figures in Haile Selassie's regime, and all of them had been overthrown along with him. No one had been cultivating the outs, who were now in charge. The embassy itself had been caught flat-footed so far as knowing anything about the new players in Addis Ababa, and Washington was almost entirely lacking in the "intelligence" needed for my contingency paper. As days passed into weeks, the imperious and imperative demands raining down on me and my office from Henry Kissinger and his associates on the seventh floor of the State Department became ever more insistent and threatening. The Secretary demanded to know just what had happened to his friend, the Emperor, what his current status was, and what was likely to happen to him. Beyond knowing that he was under some form of house arrest and certainly had no chance of regaining power, all we could offer was speculation.

The CIA was not solely responsible for this intelligence failure in Ethiopia, but it was the agency most responsible for knowing that there existed a group of disgruntled, leftist officers in the Ethiopian military who were capable of plotting the overthrow of the Emperor and the installation of a radical, Marxist, anti-Western regime close to Moscow.

Greece

Greece in the mid-1960s was suffering from an instability in its domestic politics, a departure from the strong democratic tradition it had rebuilt, partly with American assistance, in the years after World War II. The principal centers of power—the two main political parties, one liberal centrist and the other conservative; the palace; the military leadership; and the business and professional establishment—were vying for dominance with increasing rancor and a seeming inability to compromise. This turbulence was resolved—some would argue that it was exacerbated—by a military coup d'état on April 21, 1967, engineered by a bunch of right-wing, basically fascist army officers, mostly colonels, who had been plotting such a takeover for some years.

There has been speculation—outright belief in some quarters—that the 1967 coup was a CIA-initiated or -supported covert action. The fact is that it was much more of a CIA intelligence failure. The CIA, our military attachés, our palace connections, and, in fact, practically all elements of the American mission were firmly focused on a planned coup by the senior generals of the army, condoned by the palace, whose aim was principally to keep the Center Union party, led by George and Andreas Papandreou, from returning to power. We had excellent information on this plot, but it never took place, because the aforementioned colonels, aware that the generals were about to make their move, preempted that coup with their own.

I was one of two officers in the embassy who advocated a proactive policy, suggesting that we ought to try to reverse the coup in the early days and put the government back in the hands of its constitutional authorities. My anti-junta stance earned me the hostility of several CIA and military officers who were vociferously pro-junta (these were mainly Greek-Americans with extremely reactionary views) and who even stooped to labeling me "probably a communist" behind my back.

With almost all of our reporting officers of the various agencies not unhappy that we were now in bed with "the colonels," it fell to me and a very few others to keep in contact with the opposition, such as it was—mainly the leading pre-coup politicians not in detention, various intellectuals, and other opponents of the junta. My efforts reached a low point when I discovered that my CIA colleagues had spread the word around Athens that I was actually a CIA officer under the "deep cover" of a regular Foreign Service diplomat. This canard was somewhat successful in discrediting me with the democratic opposition and certainly made my reporting task much more difficult, as some of my sources dried up and my overall credibility suffered.

Here was a classic case of our clandestine operatives working in league with the ins, forcing our open diplomats to work in clandestine ways to keep in touch with the opposition—an opposition that was sure to return to power when democracy was restored. Most of the CIA's reporting at that time served the useless function of trying to make the junta look good, softening its image, deceiving our leaders about its intentions to restore constitutional government, and discounting its abominable practices, such as the torture of political prisoners and other dissidents.

This biased reporting had serious consequences. It fell to me to give my Ambassador a month's warning of the King's attempt on December 13, 1967, to overthrow the colonels—a report based on information received from opposition sources. My superiors did not take this report at all seriously, and the CIA completely ignored it. The King's effort failed utterly, having won no support from us or anyone else, and he went into exile, where he remains to this day. This was the last chance to dislodge the colonels before they became entrenched. What else was this but another "intelligence failure"?

When I was reassigned to Athens fifteen years later, a great many things had changed. Democracy had been restored and the domestic political competition was vigorous. Andreas Papandreou, the target of the junta in the 1967 coup, was now the leading political actor in Greece, head of a socialist party he had founded, and he had just won a second four-year term in office through hotly contested elections. The leaders of the junta were serving life prison terms. The U.S. government had been working assiduously to repair the damage to our relations with Greece that had resulted from our seven-year toleration of the junta and our alleged failure to stop the Turks from invading Cyprus in 1974. We were facing the arduous task of renewing the agreements authorizing us to keep four military bases on Greek soil, a relationship that had become problematic, even though Greece was a member of NATO, because of widespread anti–U.S.-government sentiments among the Greek people.

Also, I was now the Ambassador and my relations with the CIA station in our embassy took on a distinctly different character. Put simply, we cooperated successfully in a number of important areas. I will cite only a few examples, as I believe they illustrate the Agency's potential usefulness in operations as well as intelligence.

The CIA was, as usual, obsessed with recruiting Soviet and other communist defectors, and we actually obtained one, a senior GRU official who was a "walk-in" and who complicated his defection by abandoning his wife and insisting on bringing with him his son and the son's tutor, who was also the man's mistress. He was a prize catch, and the CIA successfully extracted him to Germany and then the United States using available military assets. The Station Chief kept me fully briefed on developments throughout this operation, and at one point I had to play an important role in advising the Prime Minister privately of what was afoot so that he wouldn't react inappropriately and publicly when it was later revealed that the Soviet official had defected from the Soviet embassy in Athens. The local newspapers speculated greatly about why I had called on the Prime Minister at his residence at 11 p.m., but they never were able to pin down what the unusual encounter had been about.

The CIA also kept me fully briefed on another of its important covert operations, lest I be called upon to explain to our host government what we were up to, a precaution that in this case proved unnecessary. To capture a notorious terrorist so that he could be put on trial in the United States for the murder of Americans, the CIA mounted a sort of sting operation, luring the terrorist to Cyprus to receive a big payoff for a drug-distribution deal. The terrorist ended up boarding a yacht that the CIA had chartered in Greece; it sailed to international waters off Cyprus, where the man was arrested by FBI agents and transported to the United States. The only connection with Greece was the yacht charter, and that never became an issue. This kidnapping on the high seas was of questionable legality, but was justified on the basis that terrorists obey no laws and so are not entitled to legal protections anywhere.

Matching the CIA's obsession with recruiting communist defectors was my own obsession with countering the threat the American mission faced in Athens from a domestic anarchist-terrorist organization that had murdered several of our military and diplomatic officers during the previous decade. My obsession became paramount after our defense attaché, Navy Captain Bill Nordeen, was blown up by a car bomb a block and a half from his home as he was driving to work one morning. I decided we could no longer rely on our host government to be our principal protection against this terrorist threat, and that we would have to take matters into our own hands.

When my pleas to the State Department were answered with lame complaints about a lack of resources to mount such a sustained effort, I turned to the CIA, which was happy to oblige, for its officers were prime targets of this terrorist group; the first victim, in fact, had been Station Chief Richard Welch. The CIA had the resources, mainly specially skilled personnel, and these were provided with alacrity. We had few successes by the time my tour in Athens ended, but that was not for want of trying. In one instance, the CIA's experts were able to unmask a local fabricator who claimed he could provide important information about this terrorist group but who failed a polygraph test miserably.

The cooperation I experienced from the CIA in this realm reassured me that perhaps many of the problems of the CIA arise from its being asked to do things that it—that no one—ought to be asked to do, and that the blame should therefore most often be placed on the taskmasters rather than on those instructed to do the improper or the impossible.

Recommendations

1. The most basic change should be to convert the CIA into a still independent but fundamentally analytical intelligence agency, whose primary function would be to prepare objective, nonpoliticized reports and estimates for policymakers in all branches and departments of government, using the information gathered from all available sources, open and clandestine. Responsibility for covert actions, to the very limited extent still deemed necessary and approved by the President, should be turned over to the Pentagon.
2. The primary providers of political, economic, and security intelligence from overseas, almost all available from open sources, should be the professional diplomatic officers of the Foreign Service. The resources devoted to this effort should be greatly increased (as CIA appropriations are reduced) to provide more Foreign Service personnel and to improve the Foreign Service's capabilities in foreign languages and regional expertise.
3. To the degree they are still needed, clandestine intelligence collectors and reporters should concentrate on intelligence that is obtainable only clandestinely—for example, on terrorism, the proliferation of weapons of mass destruction, and international criminal gangs and conspiracies. Though I consider our "war on drugs" hopelessly misguided, useless, wasteful, and misdirected (away from the demand side and treatment of addicts), to the degree that we still consider combating narcotics trafficking overseas to be worthwhile, that activity should be left to the DEA. The few clandestine collectors still needed overseas could continue to be housed and protected in our embassies and consulates.
4. Ambassadors should be consulted about proposed covert actions in or affecting their country of accreditation while those actions are still under consideration, as Ambassadors definitely have a "need to know" in order to carry out their responsibilities properly. Their views should be weighed in Washington in advance.
5. Relations with foreign governments and regimes should never be primarily in the hands or channels of clandestine operatives. This is especially true for nondemocratic countries. All official relations with all countries should be handled by regular diplomats. Where appropriate, contact with and collection about opposition groups, especially if outlawed, can sometimes best be handled by clandestine operatives and their agents.

NOTES

1. "Spooks" is slang for spies. "Cowboys" is a term of derision for out-of-control covert operatives whose love of intrigue drowns their common sense. "Cookiepushers in striped pants" is a long-standing derisive stereotype of professional diplomats, often cited by isolationists convinced that they are effete, elite snobs affecting foreign mannerisms and disdainful of robust American culture.

2. This presidential instruction was codified in law during the administration of President Ford in the State Department/USIA Authorization Act for fiscal year 1975 (Public Law 93-475). It reiterated the Ambassador's authority over all U.S. government personnel, activities, and operations, and required such personnel to comply fully with his directives. It also required home departments and agencies to keep the Ambassador "fully and currently informed with respect to all activities and operations" of their personnel in the country of accreditation, and to insure their full compliance with the Ambassador's directives.

3. As of May 1998, the United States had 162 embassies abroad, two branch offices of embassies, eleven missions (e.g., to international organizations), one interest section, one liaison office, sixty-two consulates general, and twelve consulates, making a total of 251 diplomatic/consular establishments manned by American diplomats. There were also thirty-seven consular agencies staffed by foreign nationals. In twelve additional countries, the United States has Ambassadors accredited but no physical presence. That makes a total of 174 states with which the United States has normal diplomatic relations. Before World War II, the United States had relations with around sixty-four countries.

4. Largely as a result of the 50 percent cut in financial resources, in the past five years the United States "has closed or scheduled to close 36 diplomatic or consular missions, 10 USIA posts, and 28 USAID missions" (quoted from Harry C. Blaney III, "Diplomacy's Bad Deal," *Foreign Service Journal,* November 1997: 43–44).

JACK A. BLUM

5 Covert Operations
The Blowback Problem

WHEN THE SUBJECT of the "blowback" of covert operations arises—that is, the damage that such operations do to this country—most people think of the obvious examples, such as the World Trade Center bombing by Afghan war veterans or the Watergate burglary by Bay of Pigs veterans. Some might focus on American assistance to former Nazis and war criminals, enabling them to influence domestic and international politics. But the most serious blowback of America's covert activity has been its contribution to the steady erosion of important elements of constitutional democracy.

The United States Constitution is at the heart of the blowback debate. To state the issue plainly: Are covert operations compatible with our constitutional form of government? If, to protect the national security during the Cold War, we took risks with the Constitution, should we continue to take the risks now that the Cold War has ended? Are there less damaging ways to reach the same goals? Would a different approach to managing covert operations create fewer problems at home? The need for a discussion of these fundamentals is urgent, yet instead of discussion, there has been silence.

The assumption that an informed electorate will hold elected officials accountable for their conduct is at the heart of our government. In the period between elections, we make the further assumption that because coequal branches of government share power, there will be checks and balances. We assume we will be protected by the fact that no one branch of government will allow another to get away with anything outrageous.

Events have taught us that covert action is an exception to these assumptions. Requirements of operational secrecy and the need to protect "sources and methods" have repeatedly defeated public accountability. The attempt at developing effective congressional oversight that came out of the Church Committee investigation (the final report was issued in 1976) failed, and with the failure, the system of checks and balances stopped working.

Although the Constitution does not require the United States to have an official or public history, accurate, recorded history is one of the most important protections against corruption. History forces responsibility and accountability. Access to the historical record is the beginning point for serious political debate and action.[1] The secrecy surrounding covert operations and the scope of covert activity have produced a history that is missing important chapters and key pages.

The end of the Cold War offers a chance to assess the constitutional impact of covert operations on democratic institutions, to evaluate and balance the pros and cons. Did the positive achievements during the Cold War justify the harm? Is the problem of damage to democratic institutions inherent in covert operations, or is it the particular product of the permanent covert-operations bureaucracy developed in the Cold War?

A real debate over these issues will lead the United States to some of the same arguments that other countries are struggling with. The Czech Republic started a process it called *Lustrace:* a public cleansing to enable the country to deal with the problems that its system of secret police and secret informers had created. In South Africa the process has been run by the Truth and Reconciliation Commission, which has put on the public record the secret action on all sides of South Africa's struggles. Germany has been forced to face its Nazi past and its Stasi past. France has been forced to deal with issues of collaboration with the Nazis. Argentina is wrestling with the issue of responsibility for the "disappeared."

What did we do in secret to win the Cold War? Can we reconcile that past with our hopes for the future and with an effort to reinvigorate democratic institutions? Can there be a new foreign policy without confronting the past?

The large gap in our history books—the secret steps we took to "win" the Cold War by manipulating other governments, international organizations, institutions, and political systems—should be filled. Perhaps with our missing history in hand, we can begin to formulate answers to today's pressing international problems, such as corruption, organized crime, terrorism, and arms trafficking—many of which have connections to our secret past.

BLOWBACK PROBLEM 1: DISTRUST

Everyone, both here and abroad, knows there has been wide-ranging American covert activity. Operations and suspicions of operations have become public in almost every corner of the world. Some operations, such as the attempts to assassinate Fidel Castro, have been fully exposed by congressional hearings. Some have been so large—Afghanistan and Vietnam—that they never were secret. Others have been secret in the United States but well known to their targets. The Soviets knew all about

American operations in the former Soviet Union because of Aldrich Ames. Very few Americans had the same information.

Dozens of well-researched nonfiction books have detailed past covert operations.[2] Still more have been the subjects of fictional accounts that have thinly disguised the underlying facts.[3] And a vast outpouring of spy novels and films has amplified the sense that the hidden hand of government reaches into every corner of domestic and international life.[4]

The perception of an omnipresent CIA is damaging to democratic discourse. It has made the motives of participants in public debates suspect. A substantial number of American citizens are convinced that there are covert plots underneath the most ordinary events. Many otherwise rational people believe that some of their fellow citizens are under the direction and control of the intelligence services and that they participate in public life as part of a secret agenda. The militias, the far right, and the far left all hold these views. This perception of covert manipulation of domestic political affairs will be very difficult to change. Yet changing this widespread paranoid vision is essential to restoring a sense of citizen ownership of and participation in government.

The existence of a large, permanent intelligence bureaucracy engaged in secret activity feeds the distrust. The laws that limit CIA activity to foreign locations and against foreign targets have not convinced any of the skeptics that operations on American soil, aimed at domestic targets, are not in progress.

This distrust has been building for decades, and has been fueled by repeated stupid excesses and improprieties in domestic activities. The catalog of covert-intelligence idiocy makes for painful reading. It includes testing LSD on unsuspecting subjects, resettling Nazis in the United States, and funding right-wing anticommunist exiles in this country.[5] Some of the more benign activities have included funding intellectual literary magazines, supporting a publishing house, and financing friendly domestic student movements.

The tumultuous events of the 1960s had everyone, from Presidents on down, wondering what was really going on behind the scenes. On recently released tape recordings, both John F. Kennedy and Lyndon Johnson express skepticism that they are being fully informed about what the covert operators are doing. For years, polling has shown that the vast majority of the American people believe that there were conspiracies behind the assassinations of the Kennedy brothers and Martin Luther King. Most of the theories advanced publicly involved the CIA and the FBI. The belief in these conspiracies has persisted despite congressional reinvestigations of the assassinations and the declassification of millions of assassination-related documents.

The central problem with disclosing full information about those events at the time was the need to cover truly stupid covert activity. The investigation of the Kennedy assassination quickly led to disclosures about CIA attempts to assassinate Castro in cooperation with organized-crime figures. The King investigation led to the discovery that the FBI had been running covert surveillance operations on King, had run a disinformation campaign against him, and for years had been attempting to disrupt his organization's activities.

During the Watergate affair, President Nixon used the CIA and the FBI, and then dragged them into his efforts at a cover-up. He tried to portray the operations of the "plumbers" unit at the White House as related to national security. As the investigations progressed toward impeachment, Nixon's partisans turned the tables and began to suggest that the covert operators in the intelligence services were trying to get him out of office. The Watergate investigation, and the abuses it unearthed, further eroded public confidence in the accountability of government for its covert works.

The erosion continued in the Reagan administration. Reagan's national-security team used covert-operation techniques on the American electorate. It bypassed the legal processes for control of covert operations by turning the National Security Council from an advisory to an operational agency. It covertly violated export-control laws by shipping arms to Iran and circumvented legislation on operations in Central America by funneling money through American client states.

Because so many of the truly incredible tales of the past have turned out to be true, many Americans have reached the point at which pure fables are taken as credible and worthy of investigation. When kooks announced that Vincent Foster, an aide to President Clinton, had been murdered to cover an exotic international conspiracy in the White House, much of the public was prepared to believe that there must be something worth looking into. The Internet is home to numerous stories about covert conspiracies involving President Clinton and a secret base at Mena, Arkansas. These stories belong in the category of science fiction, but the lies cannot be put to rest because so many previous denials about covert fiascos have turned out to be wrong. The history of covert operations and public denials has blown back to contaminate our domestic political space and provide a genuine basis for suspicion.

The distrust of the covert world is not limited to the general public. Members of Congress have publicly expressed frustration and disgust with the deliberate manipulation of public opinion by the intelligence community, whether through lies or evasions about operations or through the distortion of intelligence reports. The members of Congress who wrestled with the issue of impeaching President Clinton for perjury would have done well to consider the artful, and not so artful, dodging of questions by intelligence officials whenever they are questioned about embarrassing or illegal operations.[6] Not a word of their testimony can be taken literally. Phrases must be parsed and technical definitions considered. For example, when the CIA is asked whether it has a file on a certain person, the answer can be "No" even if the file exists, because the file is kept under a pseudonym, and the records that associate individuals with their pseudonyms are not kept in files.

Members of Congress have feared taking on the intelligence establishment because they know it fights back. Consider the vicious attacks on Congressman Robert Torricelli for his disclosure of the relationship between American intelligence operatives and murderers in Guatemala. When Senator John Kerry questioned the Contra war, John Hull, then by his own account on the payroll of the National Security Council, held a press conference in San José, Costa Rica, and called Kerry a communist.

BLOWBACK PROBLEM 2: COVERT TECHNIQUES AS PART OF THE AMERICAN POLITICAL PROCESS

American covert operations during the Cold War developed the manipulation of democratic political systems into a science. In some operations, such as the efforts to keep the communists out of power in Italy after World War II, the United States covertly funded political parties and friendly politicians. In others, such as the overthrow of the Arbenz government in Guatemala, we used disinformation to destabilize an elected leader. In Poland, to push an emerging democracy our way, we supported the elements of the society that would be most inclined to support American interests: Solidarity and the Catholic Church.

In the process of developing these techniques of intervention, a cadre of people learned the art of the political dirty trick. One—Richard Nixon—was a particularly able student. The 1972 presidential campaign, with Nixon's direct involvement,[7] put the techniques to work. Political money was laundered through Mexico, and disinformation—campaign sabotage—was used to destroy the candidacy of Edmund Muskie.

But American leaders were not the only ones to use these techniques to damage our political system. Foreign leaders who worked with us, or who had lost because of what we did in their countries, were also apt students. They realized that American elections were subject to manipulation, and the open process in American politics made them an ideal target.

The Nationalist Chinese were the first to make use of the covert campaign contribution as a tool to insure American political support. In the apartheid days of South Africa, the South African government spent millions on agents who tried to buy newspapers and plant stories favorable to their government. The Korean Central Intelligence Agency has long operated in the United States, working through various vehicles to support conservative causes. The government of India was caught in a scheme in which an Indian-American fund-raiser in Baltimore reimbursed "contributors" to designated congressional races. In the most recent elections, attention focused on the governments of China and Indonesia, both of which appear to have put money in the campaign coffers of President Clinton.

As the importance of the United States in world affairs increases, we have to assume that foreign governments will continue to use our own techniques against us. The question is whether we can continue to intervene at will elsewhere and at the same time protect the integrity of our own processes.

BLOWBACK PROBLEM 3: COVERT OPERATIONS IN DOMESTIC POLITICS

In at least one case a covert operation was directed at the United States by its own agencies. This happened in the 1980s, when the Reagan administration orchestrated a campaign of "public diplomacy" directed at Congress and the American people. The campaign was designed to produce public support for the war in Central America. It included giving the Contras assistance in lobbying Congress, picking their leaders based on their ability to lobby, and encouraging them to attack members of Congress who opposed the Reagan policy.

Oliver North, running covert operations from the National Security Council, interfered with American law-enforcement processes, leaked sensitive intelligence to promote the Contra cause in Congress, and worked with the Contras to develop news events in the United States that would put them in a favorable light. The CIA wrote the Contras' congressional testimony, scheduled their appointments, and helped them manage their public relations. The American people were the target of that operation.

BLOWBACK PROBLEM 4: COMMAND AND CONTROL

When secret operations and secret agencies abound, the issue of command and control is critical. Are the operations pursuant to lawful orders? Are they in line with national objectives, or are they designed to protect the bureaucracy from political control? Can elected officials control a covert operation's bureaucracy, or will the bureaucracy take charge? If command and control of covert action fail, the potential for constitutional mischief is extraordinary.

In recent years, there have been times when the issue of control has been in doubt. It has been the nation's great good fortune that, with very rare exceptions, none of the intelligence activities that have provoked public outcry has been the result of an independent frolic. The problem has been that the operations were designed to provide the political leadership with plausible deniability. The object is to allow political leaders to blame the civil servants. This makes the issue of determining legitimacy very difficult.

The most important check on the legitimacy of covert activity is provided by counterintelligence. Counterintelligence is, in theory, supposed to detect operations being run by adversaries, and its attention would thus be drawn to any covert activity that was not part of the official program. Early in William Casey's tenure at the CIA, he undercut the role of counterintelligence, a move that freed his irregular Central American operatives from the controls built into the system.

The culture of secrecy empowers cowboys. Until recently, they could develop covert relationships with potential assets tainted by drug and human-rights problems—as they did in Haiti and Central America—without approval. They could establish working relationships with extreme military and political groups—as they did in El Salvador and Argentina—without supervision. They can, if not properly controlled, also use their skills to influence the political climate in which they are working by selectively leaking information.

The loss of control of the intelligence community's "black budget" has been a serious problem. In 1996 Congress discovered that the National Reconnaissance Office had an unaccounted surplus running into the billions of dollars. The NRO has used some $300 million of that to build a state-of-the-art headquarters. Because of the secrecy surrounding the budget, the normal contracting controls could not be used. Issues of wasteful spending and improper relationships with contractors could not be raised. There are no General Accounting Office audits and no contract challenges.

When operations are run through commercial fronts and make a profit, the question of control over the operation and the money is an issue. Congress took steps to end the accumulation of profits in proprietary companies, but the companies are difficult to control. Operational involvement with corrupt banks, such as BCCI, the Castle Bank, Nyguen Hand Bank, and the Bishop Rewald brokerage firm, have raised serious questions about the control and management of funds.

When efforts are made to investigate possible inappropriate activity, the investigation founders because of secrecy. When the Banco Lavoro/Iraq loan scandal broke, the Attorney General ordered a full Justice Department investigation. When the report on the investigation was issued, it had an attached appendix, at first classified but then declassified. The appendix explained that the investigation of the Department of Justice was inconclusive because Justice could not be sure it had seen all the relevant documentation. The Justice report said that even though its investigators had the full cooperation of the CIA general counsel, the investigation had led to parts of the CIA that the general counsel did not know existed, and to filing systems that were deliberately organized to thwart investigation. If the Attorney General, with the full cooperation of the CIA management, cannot reach the truth, can anyone?

On a more subtle level, there is the problem of educating American Presidents so that they have the capacity to evaluate and control their own intelligence apparatus. Very few Presidents come to office prepared to deal with covert-operations issues. Without independent information and preparation, they can be manipulated. Unfortunately, the problem has never really been addressed within the government, much less solved.

Given these blowback problems, in addition to the more obvious ones, why has the steady erosion of democracy that accompanies the covert-operations approach to foreign policy been so readily accepted by so many political leaders? The answer is that going covert is easy. It is the path of least resistance and requires the least in the way of real leadership. Consider the massive effort George Bush had to make to mobilize the United States and the rest of the world to deal with the invasion of Kuwait. Congress was recalcitrant and other members of the alliance had to deal with a variety of pressures from their own people.

We Americans like quick, painless, inexpensive fixes to complex problems. We especially like them in foreign policy. For a President faced with a foreign crisis, a covert operation that promises quick and painless results is very seductive. Risks seem small, deniability is built in, and failure will be kept secret. Covert operations also appeal to the romantic and adventurous side of Americans. James Bond, Tom Clancy's heroes, and a host of other stars of fiction and film have reinforced the positive view of the spies who manipulate governments. Presidents know that successful covert operations are never a domestic political liability.

In general, the voters in democracies hate war. The debates that have preceded overt American military intervention in foreign lands have invariably been intense and bitter—the Gulf War debate and the debates over Vietnam are but two examples. Covert operations—intervention short of war—give Presidents freedom from these difficult and annoying democratic constraints. Although Congress must be notified, the notification is limited to the Intelligence Committees and leadership. In some cases Congress can be told of operations after the fact. When the information is conveyed, it is in general terms without significant detail. The briefings themselves are under constraints of secrecy so draconian that, once briefed, members of Congress are unable to act on what they have learned.

For Congress, the attraction of this system of limited notification and no public debate is that it relieves the members of responsibility for making a tough decision. For the White House, the limited notification makes the Congress a willing accomplice. Congress has learned that knowledge is responsibility and that not knowing is the easy way out.

BLOWBACK PROBLEM 5: DISPOSAL

Cuba

No discussion of the blowback problem can be complete without a look at the "disposal" issue. Assets and operatives do not disappear. Unlike in the movies, a covert operation does not end when the lights go on. The operatives, their agents, their subagents, and their training hang around for years, sometimes decades. Few of them would fit into what Alexis de Tocqueville once called the United States: a school for democracy.

The efforts to depose Fidel Castro as the head of the Cuban government are very public. The "blowback" from those operations is a handbook for students of the disposal problem. Following the defeat at the Bay of Pigs, the American-trained and -sponsored invasion force was imprisoned in Cuba. Its release was negotiated by the Kennedy administration and the veterans came to the United States as political refugees. As part of the settlement with the Soviet Union at the time of the Cuban missile crisis, the United States secretly agreed not to invade or overthrow the Cuban government. This promise left a lot of very frustrated Bay of Pigs veterans looking for an anti-Castro outlet.

In an effort to keep the veterans busy, the CIA set up "Operation JM/WAVE." It became the largest employer of Cuban Bay of Pigs veterans and one of the larger employers in the Miami area. It was much like the exercise wheel in a gerbil cage. The idea was to keep the veterans busy and give them the illusion of activity, but to prevent real trouble.

The JM/WAVE participants launched a series of inane assaults on the Cuban government from American soil. These failed, as they had been expected to do. But, probably without CIA approval, the Cuban exiles also aimed a series of attacks on domestic targets. These included planting bombs at businesses that had Cuban connections and businesses that shipped packages to Cuba through third countries.

One of the most absurd but dangerous of the exile attacks was the Novo Sampol brothers' attempt in the late '60s to shell United Nations headquarters with a bazooka from a parking lot in Flushing, New York.

Fortunately, the shells they fired fell harmlessly into the East River. The New York police arrested the Novo Sampol brothers, but they were soon bailed out and released to return to their homes.

Over the years, the scope of the Cuban exiles' targets widened. Bombs were put in the cars and homes of those who spoke out against the most extreme exiles' view of U.S. relations with Cuba. One car bomb blew off the legs of a Miami radio talk-show host. In 1986, a pipe bomb blew out the wall of a Miami museum that had had the temerity to display the works of an artist who had not properly denounced the Cuban government. The attacks supported a campaign of terror and intimidation in Miami directed against anyone who was not sufficiently anti-Castro. At different times the campaign targeted the *Miami Herald,* performers who had traveled to Cuba, athletes who had participated in games there, various moderate organizations, and outspoken independent journalists.

The exile groups became a pool of talent available for "dirty tricks" directed against communists and "leftists" throughout the hemisphere. They were used by the CIA as "unilaterally controlled Latin assets" to assist in a range of covert projects. Cuban exiles turned up in Chile, Bolivia, Argentina, and Venezuela, and throughout the Caribbean and Central America, as assistants in anti-leftist causes. Some were official government actions and some were not. The exiles did not seem to know or care which were which.

But the Cuban talent pool was also available for domestic work. The Watergate burglary team was recruited from this group. In 1976, the same pool of exiles provided help for the Chilean government of General Pinochet in the murder of Orlando Letelier. Letelier had been Salvador Allende's foreign minister. The murder was carried out in downtown Washington with a remote-controlled bomb that had been placed under the front seat of Letelier's car. It was detonated near Sheridan Circle, killing Letelier and his assistant, Ronnie Moffit. The Novo Sampol brothers participated in that effort.[8]

When the war in Nicaragua began to take form in the early 1980s, the exile community continued to provide "talent" and funding. The exiles, some of whom had drug-trafficking connections, played a central role in setting up the Contra operations. Before the United States played an active part in assisting the Contras, the Cuban exiles worked with the Argentine government and with various "private" groups in the United States to ship weapons to Central America and to train the Contras.

Because of this covert-operations talent pool, Miami became a major center for arms dealing, money laundering, and revolution. Residents joked that Miami had become America's Casablanca. The Bay of Pigs may have failed to push Castro from power, but its follow-up, JM/WAVE, and its exile leftovers have permanently altered Miami's and America's political landscape.

Afghanistan

The blowback from the Afghan war illustrates a different kind of disposal problem. If the issue with the Bay of Pigs veterans arose because they moved to Miami, the issues involving Afghan veterans arose because of globalization. And the Afghan disposal problems seem likely to be with us for years to come. (See Chapter 2.)

When the Afghan war began, American planners were determined to avoid a repeat of Vietnam. There would be American support for the resistance, but no American troops and no American presence. To mask American participation, support for the Afghan resistance was funneled through the Pakistani military and the Pakistani intelligence services. To help control the costs and avoid congressional complaints, the United States encouraged the conservative governments of the Persian Gulf, especially the Saudis, to help fund the resistance. They, in turn, made the fund-raising politically acceptable by characterizing the effort as a jihad against godless communism.

In the early stages of the war, the American strategy was to let the Russians learn the lesson that the British learned in the middle of the nineteenth century: Invading Afghanistan is a losing proposition. The Afghans have been destroying invaders for centuries. They perfected guerrilla warfare and wars of attrition long before the United States was a country. Simply giving the Afghan opposition the tools and letting them fight would—or so Washington thought—be enough to seriously bleed the Russians.

As the war progressed, however, Washington began to assume a more active stance. Stinger missiles were introduced. The Saudis were encouraged to fund the recruitment of young Islamic fighters from all over North Africa to join the jihad. American covert operatives suggested that a cadre of anticommunist Islamic fighters could be developed and trained in the Afghan war and that these fighters would then return to their home communities with strong anti-Soviet feelings. The theory was that this cadre would be an important asset in the struggle with communism for the control of the Arab world. Recruiters began to work with the young unemployed in Islamic communities around the world. From the slums of Cairo to the streets of Brooklyn, they found willing volunteers. The volunteers were then indoctrinated in training camps in Pakistan. The indoctrination was in a militant version of Islam, more religious than military and more anti-Western than anticommunist. The fighters produced in these camps viewed the Afghans they were helping as not sufficiently fanatic. They showed contempt for the more relaxed Afghan version of Islam. At the time, a number of advisers to the effort tried to warn the United States of the dangers that the recruitment and training program were creating, but the warnings were brushed aside. The Afghan war was a "winner."

For the first time in more than thirty years, a "covert" paramilitary operation enjoyed bipartisan political support. Congressmen toured the battle zone. Policymakers began to take credit for the "victory." But problems were obvious as early as 1984, and as the years went by they got progressively worse.

Dissension among the seven Afghan resistance groups became public. Several groups sent emissaries to Washington to warn that the Pakistani government was favoring the radical anti-American elements. Some of the resistance groups warned that the weapons being supplied by the United States were not going to the battlefield and might very well wind up in the hands of terrorists. Congress, basking in the glow of victory, did not want to hear the news.

When the Soviets withdrew in 1988, the celebration of victory was universal. Americans were told that the war was over and that the United States had won. Few noticed that the war continued—now among the "victorious" factions. Few noticed that Pakistan and Afghanistan had become the largest producers of opium poppies and heroin in the world. (See Chapter 7.) Few noticed that the training camps for the jihad fighters had not been disbanded, that the fund-raising activity to support the religious indoctrination continued, or that the rhetoric was now very publicly anti-American.

At the same time, Pakistan began to see the continuing civil war as a useful opportunity. The war kept the fanatic religious elements in Pakistan busy and at the same time gave Pakistan the opportunity to control Afghanistan. It provided an outlet for Pakistani manufactured arms and insured a continuing flow of Saudi money.

The training camps and the religious training centers were now producing virulently anti-American fighters who were ready and eager to martyr themselves for the cause. The trainees and the veterans became the force behind anti-Western terrorism in Algeria, Egypt, France, and the United States. The World Trade Center bombing was blowback from the Afghan war, as were the bombings of American embassies in Kenya and Tanzania.

In the 1990s, it became obvious that the veterans of covert war could be a serious problem even if they were not resettled in the United States. Even when the covert operation was on the other side of the globe, as it was in Afghanistan, progress in transportation and communication gave the operatives the ability to strike at American targets anywhere in the world, including in the United States itself.

Blowback Problem 6: Empowering the Underworld

What covert-operations people do is illegal in the country they are working in. They are spies, and in much of world they would face the death penalty if caught. To be effective, a good covert operative must work with people who know how to get around the rules. These are people who know how to move money without a trace, can smuggle people and goods in and out without clearing customs, and have bribed or suborned key government officials. The people that fit this definition are career criminals.

Time and again, intelligence operatives with important missions have turned to the underworld for help. In Japan, the Yakuza helped the United States keep leftist labor organizations under control. The French-organized crime groups helped keep leftist labor from controlling the port of Marseilles in the postwar period. The Italian Mafia helped the United States invade Sicily and helped keep Italy from going communist. In the collapsing Soviet Union, it now seems clear, we used organized-crime figures to get information and to assist our operatives.

The most bizarre connections with the criminal underworld came during the Central American war. In testimony before the Senate Foreign Relations Committee, Leandro Sanchez-Reisse, a civilian employee of Argentine military intelligence, described how he received intelligence assistance in setting up a money-laundering operation in Fort Lauderdale. The money was to be supplied to the undercover battalion of the Argentine military that ran operations against communists and leftists throughout the hemisphere. Sanchez-Reisse also described the connections between the Argentine military and the "cocaine coup" in Bolivia. He told the committee that cocaine funds had financed ongoing anticommunist operations. His story was one of many similar accounts. According to the CIA's Inspector General, at one point the Agency employed a Honduran drug smuggler to assist it with logistics because there were no "acceptable alternatives" in the region. The Ollie North network employed a variety of individuals with drug- and gun-dealing connections to establish its supply networks.

Official intelligence dealings with criminals, especially criminals who are involved in activity on U.S. territory, raise very difficult questions. Who performs the cost-benefit analysis? Who has the authority to waive the enforcement of criminal laws? Does the government have responsibility for the effect of the sanctioned criminal activity? During the Central American war, these questions were never properly debated.

The same questions are on the table when law-enforcement agencies run undercover operations. Over the years, law enforcement has developed very strict rules to insure that the operations do not get out of control and do not do more harm than good. Despite these controls, operations sometimes go wrong. Techniques such as controlled deliveries of drugs are rarely approved because of the risk that the drugs will wind up on the streets.

As long as the criminals had little or no connection with the United States, these issues were not serious problems in the eyes of the U.S. officials. With globalization, however, criminals have the capacity to target victims anywhere on earth. Increasingly they target the United States because that is where the money is. The development of globalized criminal activity should raise new questions about the use of criminals.

Apart from the risk that they will use their connections and status against the United States, the real question is the empowerment of the underworld. By working with people in covert operations, we grant them power and status. Sometimes it is in the form of formal protection—a kind of "ticket punching." More often the power comes from the fact that everyone else is aware that the people are connected to a powerful secret agency that is an arm of a superpower. Such connections are not to be trifled with.

Conclusion

A review of the sorry history of blowback suggests that at least one of the laws of physics applies to covert operations: For each operation there is an equal and opposite reaction, though it may be a delayed one. Sometimes, as in the case of Cuba, there is a combination of failure and blowback. And in a few cases, such as Afghanistan, it will be decades before a full damage assessment is possible. The blowback can be in the form of a "disposal problem," an unintended consequence, or damage to fundamental democratic institutions.

The problems cannot be solved through increased congressional oversight or more rigid rules and regulations unless there is radical reform. The process, as it is now structured, makes Congress complicit without giving it real authority or even real information. One of the ironies of the CIA Inspector General's report on the Contra drug problem is that it suggests that the staff of the House and Senate Intelligence Committees approved in advance the CIA relationships with people who had been involved in drug trafficking.

It might be possible to discourage frequent use of covert options by requiring that an independent team evaluate and critique proposed operations. The team's job would be to try to look ahead and predict the long-term consequences of using a covert solution to a vexing problem. Much like an environmental impact statement, the report would have to accompany the paperwork in which the plans are presented to the political leadership for approval.

But the constitutional issues and the damage to the American political process suggest that a much more sophisticated assessment of potential damage is required. The people who authorize covert operations should be required to use the same balancing test that the courts use when constitutional rights are curtailed, which balances the threats to life, physical safety, and security against the value of the rights. In applying the balancing test, they should be required to create a reasoned written justification that would be subject to automatic declassification. At least there would be historical accountability.

Finally there is the issue of missing history. No honest effort to evaluate and debate the worth of covert operations can take place as long as the history stays missing. The debate is reduced to an argument with insiders who say they know that the system works because they have access to classified files that the rest of us are not allowed to see. That's a hard argument to win. We need to begin an honest evaluation of the things that were done to win the Cold War and begin a debate about the future of American foreign policy.

For this author, the answer for the future lies in reestablishing the role of public diplomacy. The diplomatic service should be strengthened, its size increased, its training improved, and its mission expanded. Instead of having spies deal with the questionable elements of foreign societies, we should be engaging foreign societies on every level. The tools of the future should include more educational exchanges, cultural exchanges, commercial exchanges, and citizen diplomacy.

We Americans should have enough faith in the merits of our own democratic enterprise to believe that we can sell its benefits on those merits. We should have some faith in the ability of other people around the world to govern themselves without the United States intervening and making their mistakes for them.

Some covert operations are required—to control terrorism, keep control of weapons of mass destruction, and rescue citizens from dire circumstances, for instance. But they should be regarded as a weapon of last resort, not a convenient way to avoid political opposition.

Notes

1. Both we and the Soviets understood the importance of history as a force in politics. In the former Soviet Union, each time the official version of history changed, civil servants would go through the libraries and cut out the pages of books that had recorded the offending facts. The publication of the Twentieth

Party Congress speech by Nikita Khrushchev denouncing Stalin's crimes against his own people (1956) and the publication of Aleksandr Solzhenitsyn's *The Gulag Archipelago* (1974) were central to the eventual end of Soviet totalitarianism.

2. For example, see Audrey R. Kahin and George McT. Kahin, *Subversion as Foreign Policy: The Secret Eisenhower and Dulles Debacle in Indonesia* (New York: The New Press, 1995).

3. A wonderful example of this genre is *The Hilton Assignment*, a novelized account of an attempt to overthrow Muammar al-Qaddafi, which failed because the CIA tipped off the Libyan government.

4. Authors of spy fiction are frequently assisted by covert-operations personnel. For example, John Le Carré met with a number of intelligence operatives in preparation for writing *The Tailor of Panama*.

5. See *Blowback* by Christopher Simpson (New York: Weidenfeld & Nicolson, 1988) for a full account of the CIA collaboration with Nazis and the funding of the far-right ethnic lobbies.

6. The CIA Inspector General's Report on the Contras and Drugs, Part II, opens with a jesuitical discussion of the definition of who is a CIA "employee." The reason for the complicated analysis is to exculpate the employee for evading an obligation to report illegal activities. The CIA language makes President Clinton's perjury defense look positively logical.

7. During the Eisenhower administration, Nixon headed the "40 Committee," the group that planned and approved covert operations. In his 40 Committee role, Nixon was exposed to the techniques that our covert operators used to influence election results around the world.

8. In 1979, Guillermo Novo was sentenced to life in prison in a maximum-security institution. Ignacio, his brother, was convicted of perjury and misprision of a felony and was sentenced to eight years. For a full account, see John Dinges and Saul Landau's *Assassination on Embassy Row* (New York: Pantheon Books, 1980).

KATE DOYLE

6   The End of Secrecy
U.S. National Security and the New Openness Movement

SECRETS AND LIES have always been endemic to the functions of state. And in a democracy, public tolerance of official secrecy tends to shift, obligingly, with the tides: In times of national emergency, such as war or civil unrest, the body politic is often willing to forgo open governance in exchange for safety; in peace, the citizenry becomes assertive once again, reclaiming its right to full and informed participation.

During the long, dark winter of America's Cold War, this traditional compact between the governors and the governed fell apart. A system of secrecy first devised in the crucible of the Second World War was not dismantled after the troops came home. It took root and grew. It reached beyond the corridors of power in Washington to taint government operations across the country and around the globe. It served to hide not only the individual misdeeds and misadventures of successive administrations, but the rationale behind them. Presidents, policymakers, and legislators used the advent of the national-security state as an excuse for their evasiveness. They assumed that they could abrogate the people's "right to know" without prior consultation, just as if the United States were engaged in an open armed conflict, and that U.S. citizens would simply accept this as the price of victory. They were right, to a degree. Fearful of the prospects of a nuclear face-off, most Americans allowed the erosion of freedoms and of the openness they had once taken for granted. As a result, secrecy spread its shadow over the crafting of foreign policies, the building of weapons, the birth of entire government agencies, the spending of federal funds, and, inevitably, the play of public debate.

There were, of course, conscientious objectors. A handful of journalists, members of Congress, civil-liberties lawyers, and others rejected the new secrecy regime as unconstitutional and undemocratic. They demanded information, gathered testimony, and filed lawsuits in an effort to pierce the veil that separated the public from its government. "Free people are, of necessity, informed; uninformed people can never be free," declared Senator Edward V. Long in 1963, during one of the many hearings that led to passage of the Freedom of Information Act.[1] What safeguards remain we owe, in large part, to them.

As the Cold War came to an end, however, an extraordinary thing happened. Openness became a cause, accountability a movement. Where once a small, dedicated band of "openness advocates" pounded on the closed door of the national-security state, now crowds appeared, an astonishing chorus of voices that joined the call for the end of secrecy: librarians and archivists, academics and historians, Republicans and Democrats, scientists, jurists, human-rights activists, and members of the defense and intelligence establishment. This new and unusual constituency, forged in the wake of half a century of covert operations, black budgets, and information controls, should not be confused with a "coalition," since it by no means represents uniform political goals. But it is united behind a simple and undeniable truth, most recently articulated in the opening words of the bipartisan Moynihan Commission report in 1997: "It is time for a new way of thinking about secrecy."[2] And although it has not yet managed to bring about the deep structural changes required to fundamentally transform the system, its achievements over the past decade contain crucial lessons for the future of secrecy reform in the United States.

SILLY SECRETS, DANGEROUS SECRETS

Lesson one is laughter. Perhaps the first and clearest indication of the change now under way came when outrage over excessive classification and secrecy began to mingle with ridicule.

The Central Intelligence Agency experienced this new trial-by-humiliation shortly after Robert Gates became its fifteenth director in November 1991. One of the new DCI's first acts in office was the creation of a working group to examine CIA openness policies, a response to criticism that excessive secrecy had damaged the Agency's credibility and undermined its relations with Congress. When the Task Force on Greater CIA Openness finished its deliberations four weeks later, however, its report was promptly classified, prompting a roar of laughter from reporters and editors around the country. After several months of public derision, the CIA released the report.

Countless further examples of the sheer silliness of the contemporary secrecy system abound in every corner of the government.

- In 1992, Steven Aftergood of the Federation of American Scientists, a veteran advocate of openness, filed a Freedom of Information Act request for the oldest secret record known to exist in the National Archives. The document turned out to be a 1917 Army memo about

troop movements in Europe, protected from disclosure for three-quarters of a century as sensitive national-security information.
- A decade after the collapse of the Berlin Wall, the CIA continued to insist that it could "neither confirm nor deny" whether its classified archives included biographies of long-dead communist leaders.
- The Defense Department fought a costly court battle for six years to withhold thousands of secret records on the Carter administration's failed mission to rescue the Iranian hostages in 1980. When Defense lost its case, the records released included such gems as a "Top Secret: Eyes Only" after-action report that contained a warning to pilots not to carry milk in their box lunches because it would sour in the desert heat.
- The Moynihan Commission came up with its own nominees for most absurd secret, such as weather reports produced by an aide to General Eisenhower during World War II, which, according to a former government historian who testified before the commission, were still classified thirty years after the fact.

But if ridicule has emerged as one of the new weapons in the contemporary fight for openness, the enduring power of secrecy in government is no laughing matter. Any attempt to take its true measure requires a mighty calculus. Even Steven Garfinkel, a career bureaucrat and chief of the President's Information Security Oversight Office, does not know precisely how many secret records are stored in Agency vaults. "Maybe tens, maybe hundreds of millions," he ventures. "Perhaps billions." And many more to come: Between 6 million and 7 million documents are stamped "classified" every year, about 17,000 daily. The cost of keeping so many secrets—what with salaries, safes, locks, security training, records management, computer programs, and the like—is equally staggering. Garfinkel's office figured that the government spent some $4.1 billion in 1997 alone on "security classification." That figure, of course, does not include the CIA's share, which is secret.[3]

How to explain this orgy of classification? There is, obviously, the culture of secrecy in which military and intelligence agencies are steeped, a culture that rewards obfuscation and opacity and profoundly discourages transparency. Equally important, perhaps, is the fact that there exists no penalty for overclassifying government information. Those who challenge the secrecy system, on the other hand, risk censure, sanction, or worse.

A disturbing recent example of the perils of leaking security information was the case of Richard Nuccio, a State Department official who was drummed out of government service for making what he believed was an ethical choice to inform Congress about a cover-up. In the course of a 1995 investigation, Nuccio had found classified CIA documents indicating that a paid Agency informant, a colonel in the Guatemalan army, was helping to cover up the murder of an American innkeeper and the torture and murder of the husband of an American citizen.

Nuccio went to Representative (later Senator) Robert Torricelli of New Jersey. He knew that his discovery directly contradicted senior CIA officials, who had told members of the House Intelligence Committee, of which Torricelli was a member, that their files contained nothing relevant to the case. They were wrong. But after Torricelli revealed the story to the press, the CIA stripped Nuccio of his security clearance for disclosing classified information. The decision was supported by the Clinton administration's Justice Department, and Nuccio resigned from the government in 1997. Although Congress subsequently passed a law to protect whistle-blowers from being charged with unauthorized intelligence disclosures,[4] Nuccio's fate sent a chilling message to those facing comparable situations.

Restrictive secrecy practices do more than endanger accountability. They cheat history. Despite the existence of a variety of legislative safeguards designed to protect the historical record (such as the Federal Records Act), preservation of government documents is a sometime thing, with few rules, little oversight, and powerful bureaucratic interests at stake. Currently, only about 3 percent of U.S. government records are preserved for posterity. Agencies can make unilateral decisions to "disappear" records permanently with little fear of punishment—either by deliberately destroying them or by ceasing to create them. Consider the following:

- To guarantee the secrecy of its covert "MKULTRA" program, which for twenty years ran behavior-modification experiments on unwitting human subjects, the CIA destroyed most of the documents associated with it in 1973.
- In 1974, with the public demanding increased openness and accountability in the wake of the Watergate scandal and disclosures about Vietnam, the Joint Chiefs of Staff quietly destroyed all records of its meetings going back to 1947. In 1978 it was decided to stop keeping such records altogether. The National Archives learned of the destruction in 1993.
- The Reagan White House did its best to delete its electronic-mail files during the Iran-Contra scandal and again when the administration was preparing to leave Washington in 1989. The discovery of an unknown backup collection led to a lawsuit to prevent the wholesale destruction of electronic information, and the courts have since upheld the government's duty to preserve such records.

- In 1997, a historian working with the CIA revealed that many of the original files documenting the Agency's 1953 covert operation in Iran had been turned over for routine destruction and no longer existed.

In each case, an irreplaceable piece of American history disappeared forever down the black hole of secrecy.

The official remedy for this modern malaise is the Freedom of Information Act, or FOIA. Signed into law in 1966, it was intended by Congress to guarantee citizens the right of access to government records. With the exception of a brief heyday in the 1970s, however, it has been profoundly dysfunctional. Agencies do not like the FOIA. They consider it an unwelcome trespass on their prerogatives, an obstacle to their labors, and a drain on their resources. Director of Central Intelligence William Colby no doubt spoke for many of his colleagues in the federal government when he sternly told a House subcommittee in 1974, "The Central Intelligence Agency is not a public information agency."[5]

Resistance to the law has led many agencies to engage in a kind of bureaucratic civil disobedience against its requirements, thus discouraging all but the most hardy FOIA practitioners. The opposition takes several forms. Agencies' FOIA offices are understaffed, underfunded, and generally excluded from the promotional stream that is so important to bureaucracies. Even more elemental, some agencies lack the functioning filing systems necessary to retrieve records in the first place. The CIA has been described by its former director, Robert Gates, as "archivally impaired"; in fact, the Agency's rigid compartmentalization—by which information is internally segregated and granted only to those with a demonstrable "need to know"—prevents even its senior employees from having efficient access to records, never mind the public. Delays further cripple the FOIA's utility. Getting a response to a request for information can take as little as ten days or as much as ten years, making the law impractical for anyone working on a deadline. Finally, agencies have become adept at wielding the nine exemptions permitted under the FOIA as weapons against disclosure. Experience has shown that the exemptions, which were designed to protect information such as that concerning national defense and foreign policy (exemption 1) and law enforcement (exemption 7), can be used to cover up embarrassing, unpopular, or illicit activities as well.[6] FOIA requesters on the receiving end of page after page of heavily blacked-out documents and outright denials are left to wonder whether an agency is being honest.

All three branches of government bear responsibility for the failures of public access. The executive branch has clear bureaucratic interests at stake whenever the tides of openness threaten to affect its business. Furthermore, Presidents and senior policymakers have long benefited from a system that protects their papers from public scrutiny but grants them privileged access to write their own histories or personal memoirs. Congress, which has strengthened the FOIA several times since its passage, has been less effective in reforming the underlying secrecy system and very weak in holding intelligence agencies to even minimal standards of openness. And the courts routinely uphold government claims of national security, to the detriment of citizens pressing for greater access.

Such is the enduring and pervasive legacy of Cold War secrecy today. Its effects are still felt in every aspect of government pertaining to U.S. national security, foreign policy, and intelligence: Secret agencies, secret budgets, secret programs, and mountains and mountains of secret paper are with us still. Our present system of secrecy is more than just a statutory hangover from the nuclear era, an anachronism that can be wiped away by the passage of a few good laws. It has become second nature, a kind of cultural reflex that defies rationality. The CIA showed just how deep that river runs in the same report that provoked such derision among the knowing public in 1992. "Preserve the mystique," warned the Task Force on Greater CIA Openness in a list of goals it set for the Agency's public-relations machine. Speak to the press, yes; hobnob with academics; testify before Congress on a more regular basis. But above all, "Preserve the mystique."[7]

SECRECY AND INTELLIGENCE

World War II was the original spring from which secrecy flowed.[8] Special security measures were created and then refined to protect the development of the bomb; covert intelligence and paramilitary operations; the work of code-breakers; and decision-making at the highest levels of government. War also enhanced the power of the executive branch, as President Roosevelt engaged in secret diplomacy while serving as commander in chief for the immense mobilization on the European and Pacific fronts.

By war's end, the United States had emerged as the undisputed leader of the Western world, with interests, investments, and an infrastructure that reached around the globe. President Truman and his advisers set out to design a bureaucracy that could support the country's new internationalist role. Burgeoning fears about Soviet intentions and the appeal of communism fueled the administration's first major attempt at reinventing government: the National Security Act of 1947. The legislation was a direct result of lessons learned in war, and established, among other entities, a new National Security Council to coordinate political and defense policies, and the Central Intelligence Agency to function as the President's permanent intelligence service, with the aim of preventing another Pearl Harbor.

Although the birth of the NSC and the CIA was based in open law, both agencies thrived on secrecy. One of the NSC's earliest policy directives authorized covert measures to prevent the Communist Party from winning in upcoming Italian elections. Another early NSC document, NSC 4/A, directed the CIA to wage covert psychological warfare to counteract "Soviet and Soviet-inspired activities ... designed to discredit and defeat the United States." The CIA's mission and special functions were further clarified in NSC 10/2, which gave the Agency the power to conduct a vast range of clandestine operations, including political sabotage, economic intervention, and paramilitary activities, against "hostile foreign states or groups." Hundreds of such directives issued from behind the iron curtain of the National Security Council, almost every one hidden from public scrutiny. Thus administration policy on a world of issues central to American security interests was forged and executed without the interference of public comment or debate. The secrecy that surrounded NSC decisions and deliberations from its inception precluded effective oversight, and led one congressman to remark in frustration many years later that "Congress cannot react responsibly to new dictates for national policy set in operation by the executive branch behind closed doors."[9]

Inevitably, secret plans led to secret operations and a covert apparatus to execute them. The emerging consensus among the President's men—that containment of Soviet aggression was to be America's highest calling—required a level of global activism previously unimaginable. The CIA had already become indispensable by scattering its operatives and stations worldwide and by launching a series of ambitious covert actions. In 1952, the Truman administration added the National Security Agency to the black roster, a massive intelligence-gathering entity designed to intercept electronic communications worldwide, whose basic charter was classified when it was written and remained secret for decades after. Although its primary target was the Soviet Union, the NSA was also obsessed by the enemy within, and spent its first two decades spying on Americans by intercepting their telegrams and telephone conversations.[10] President Kennedy would draw the curtain tighter during his first year in office when he established the supersecret National Reconnaissance Office to build and manage the nation's spy satellites. Not only its mission but its name and its very existence were classified.[11]

The proliferation of secrets in the national-security state required some form of control. The U.S. press was no longer willing to censor itself as it had when the country was fighting a war, and it freely published what stories it could on America's defense capabilities and atomic-energy program, infuriating President Truman.[12] In 1951, Truman followed the recommendations of an NSC committee on internal security by issuing a

The End of Secrecy 99

sweeping executive order that redefined the scope of the government's classification system. The President's order, written without reference to constitutional or statutory authority, extended classification standards that had protected military information since before the war to include the records of any and all civilian agencies that had a hand in "national security" matters. The system provided the basis for government control of national-security information for more than twenty years.

Although the executive order built a new wall of secrecy around the government, intelligence agencies sought special protections. Over time, they received them. In 1950, Truman's NSC advised that "any publicity, *factual or fictional,* concerning intelligence is potentially detrimental to the effectiveness of an intelligence activity and to the national security"[13] (my italics). The CIA's founding statute, the National Security Act of 1947, already shielded much of the Agency's business from public scrutiny, prohibiting "intelligence sources and methods from unauthorized disclosure." The Central Intelligence Agency Act of 1949 extended the cloak of secrecy to the CIA's budget, legalizing the transfer of funds from any federal agency into CIA accounts with minimal congressional oversight. In the sensitive realm of code-breaking, Congress passed the so-called COMINT law in 1950, which criminalized the publicity of information "concerning the cryptographic and communication intelligence activities of the United States." And in 1959, the National Security Agency Act gave a blanket exemption to the NSA, removing even the most basic facts about its organization, function, and activities from the public sphere.[14] This postwar surge in executive controls and national-security legislation effectively immunized the incipient but fast-growing intelligence bureaucracy against any contact with the press or public.

THE SPREAD OF SECRECY

Without the legal means to compel access, the press and public were forced to rely on "the windfalls of intelligence that are produced by cyclones of controversy," as a historian once put it[15]—on those leaks, mishaps, and selective disclosures that provided the occasional glimpse into the hidden workings of government. Secrecy now veiled ever-growing and more ambitious White House policies—the toppling of a President in Guatemala in 1954, U-2 spy missions into Soviet airspace, and the failed 1961 Bay of Pigs invasion. Error and hubris unmasked some of them. Presidents lied and were found out. As revelations accumulated, the press began to write of an "invisible government," a parallel but concealed world of spooky operators and covert machinations.

It took the crisis of the Vietnam era to force a national showdown over the secrecy system. Although Richard Nixon signed the first new executive order on classification since the 1950s, designed to break the bureaucracy's hold on secrets created since the Second World War, the President's own brazen use of secrecy as a political tool outraged the American people. Official mendacity about the war and the Nixon administration's efforts to block publication of the Pentagon Papers were the first rude shocks. In response, the tectonic plates of public perception about secrecy and its potential for abuse groaned and shifted. Watergate helped focus the nation's outrage on the executive branch. Disclosures about surveillance of U.S. citizens by U.S. intelligence agencies furthered the crumbling of public acceptance. Anger at what the United States was learning about itself was reflected in the final report of the Church Committee, the Senate select body charged in 1975 with investigating U.S. intelligence activities. "What is a valid national secret?" the committee asked. "Assassination plots? The overthrow of an elected democratic government? Drug testing on unwitting American citizens? Obtaining millions of private cables? Massive domestic spying by the CIA and the military? The illegal opening of mail? Attempts by the agency of a government to blackmail a civil rights leader? These have occurred and each has been withheld from scrutiny by the public and the Congress by the label 'secret intelligence.'"[16]

The repudiation of secrecy and executive privilege took shape within Congress through the 1974 amendments to the Freedom of Information Act. The changes empowered the judiciary to challenge government secrecy claims, and narrowed the government's ability to withhold national-security information by requiring it to prove that disclosure would cause actual harm. The new amendments transformed the FOIA into a genuine tool of investigation and historical research. Document requests from reporters and members of the public soared, as did the number of lawsuits challenging agency denials. Strong public pressure helped persuade the incoming Carter administration to issue a new executive order on national-security information, one that rested on a premise of disclosure rather than denial. According to the order, documents would be presumed open unless a classification official determined that their release "reasonably could be expected to cause at least identifiable damage to the national security."[17] Records would now be subject to a "balancing test," requiring agencies to weigh the need to classify national-security information against the public's interest in its disclosure.

The reformation was dazzling, but brief. The inauguration of Ronald Reagan brought the age of transparency to an abrupt halt, as a President sworn to the destruction of the "evil empire" quickly erected new fortifications in the name of national security. Freedom of information was an early casualty. A steady stream of restrictive controls issued forth—secrecy oaths for federal employees, broad security guidelines to limit even unclassified disclosures, and constraints on government publications—with little regard for the painful lessons of the very recent past.[18] In 1982, Reagan signed his Executive Order 12356 on classification, renouncing President Carter's pledge of openness in favor of stringent secrecy standards that vastly increased the amount of material that could be legally withheld from public view. The administration's obsession with secrecy was expressed not only in its attacks on public access but in its encouragement of all the black arts: Funding for intelligence agencies soared, covert action returned with a vengeance, and the FBI's domestic spying programs were resurrected.

It was an unhealthy obsession. Excessive secrecy bred contempt for the democratic process and encouraged some around the President to believe that they were beyond the public's reach, accountable to no one. It carried the country to the brink of constitutional crisis. The Iran-Contra affair—when a clutch of Cold War zealots seized the controls of state and launched illegal programs from the basement of the White House—was the logical conclusion for the renaissance of absolute secrecy during the Reagan administration. More than a decade later, it remains a "chilling story," as Senator Daniel Inouye, chairman of the Senate committee that investigated Iran-Contra, called it; "a story of deceit and duplicity and the arrogant disregard of the rule of law.... It is an elitist vision of government that trusts no one—not the people, not the Congress and not the Cabinet. It is a vision of government operated by persons convinced they have a monopoly on truth."[19]

Playing Games With History

Today, the ideological battlefield is deserted and the clamor for change rising. Historians have been among the most insistent agitators. Frustration over the inaccessibility of domestic archives grew at the end of the Cold War when the secret files of the former communist states began to open, releasing first a trickle and then a torrent of paper chronicling the era's great flash points. The flow of Russian, East German, and Chinese government records did little to influence the custodians of official history at home, provoking indignation among American scholars. "Why are we willing to allow our own past to be secreted, shredded, and our public misinformed?" asked the historian Blanche Wiesen Cook before the annual meeting of the American Historical Association in 1990. "Can it be that we have never decided quite who we are—an imperium or a democracy, a country of law where the people have a right to know, or a brutish informational dictatorship?"[20]

There is no better way to judge a nation's faith in itself than to read its official story. How closely it hews to the truth about the past—good, bad, or ugly—is a measure of a government's confidence in its own history and in its people. The United States has published just such a story since 1861 through the valuable document series known as *Foreign Relations of the United States*. In the *FRUS* series, the State Department traditionally has provided the most comprehensive official record of U.S. diplomatic history available, reproducing the text of thousands of original government documents and issuing them publicly in bound volumes. The series is intended to serve as the definitive account of U.S. foreign policy, and as such—according to the preface that has appeared in every volume since 1925—"There may be no alteration of the text, no deletions without indicating where in the text the deletion is made, and no omission of facts that were of major importance in reaching a decision. Nothing may be omitted for the purpose of concealing or glossing over what might be regarded by some as a defect of policy."

Noble words. During the Cold War, however, the *FRUS* fell victim to the secrecy police and was gravely damaged as a result. The trouble started in the 1950s, when the need to include military and other nondiplomatic records became evident as the volumes concerning the Second World War were assembled. Editors were faced with the task of appealing to the Defense Department, the CIA, and other agencies for the clearances necessary to obtain their documents. By the 1980s, scholars and historians who used the series were complaining frequently about excessive delays in its publication and obvious deletions of information from the documentary record.

The moment of truth arrived when the *FRUS* edition on Iran, 1951–54, was issued in 1989. To the astonishment of diplomatic historians, the volume was mute on the subject of the CIA's role in the 1953 coup against Prime Minister Muhammad Mossadegh and his replacement by Shah Mohammad Reza Pahlavi. The Agency's operation in Iran was not exactly secret. Kermit Roosevelt, chief of the CIA's Middle East operations at the time, had written all about it in his memoir twenty-five years later, even quoting the newly restored Shah as he offered a toast to Roosevelt: "I owe my throne to God, my people, my army—and you!"[21] But the *FRUS* editors had simply removed the Agency from the annals, just as the Soviet guardians of Stalin's legacy had once taken their erasers to archival photographs and smudged out the politically incorrect. The sanitized history was presented with no note to flag the omission. Instead, the records actively promoted the fiction that the coup was the result of a spontaneous popular uprising, such as in this cable from Loy Henderson, who was then the U.S. Ambassador to Iran: "Unfortunately impression becoming rather widespread that in some way or other this Embassy or at least US Government has contributed with funds and technical assistance to overthrow Mosadeq.... We sincerely hope means can be found either through US Government channels or through private American news dissemination channels for American and world publics to understand that victory of Shah was result [of the] will [of the] Iranian people."22

The edition landed like a bomb in the laps of American scholars. Its most immediate result was the resignation of the chairman of the historians' committee that advised the State Department on *FRUS*. Denouncing the volume as a "fraud," Warren Cohen, then a professor at Michigan State University, quit the committee in protest in February 1990. "The State Department is playing games with history," he wrote several months later in an op-ed piece in the *New York Times*. At the heart of the problem was "an overly elaborate, costly declassification process that encourages distortion and cover-up."23 Cohen declared that he could no longer guarantee the integrity of the *Foreign Relations* series for the scholars, journalists, and diplomats who used it.

Cohen's outrage succeeded in ripping the veil off what had been, until then, a private dance between historians and bureaucrats, performed away from the prying eyes of the general public. Year after year, the State Department's advisory panel had reluctantly signed off on completed *FRUS* volumes while urging the government to rewrite its declassification rules so that a more accurate record could be produced. The panel's concerns were echoed repeatedly by individual scholars and academic associations equally troubled by the series' deterioration, but with little effect. The storm provoked by Cohen's resignation, however, led to the critical mix of pressure, public attention, and momentum needed to compel change. A handful of senior members of Congress, Democrats and Republicans, took up the cause, proposing legislation to overhaul the *FRUS* series. Historians quickly organized an effective campaign in support of the bill and worked to advise congressional staff and lobby their representatives. Despite strong opposition from the State Department and the White House, the law passed in October 1991.24 It granted statutory authority to the historians' advisory committee and gave members the right to security clearances so they could actually read still-classified records; it required the State Department to declassify its thirty-year-old documents using standards that were less restrictive than the executive order then governing secrecy; and it ordered all agencies that participated in foreign policy to provide the editors of the series with "full and complete access" to records.

Congressional intervention did not solve all the problems plaguing the *FRUS* series. In the years since President George Bush signed the bill into law, the reconstituted Advisory Committee on Historical Diplomatic Documentation has clashed repeatedly with the CIA, wrestling over records that chronicle covert missions in such places as Indonesia, Korea, Japan, British Guiana, and the Congo. At first, disputes over withheld documents were settled by notifying *FRUS* readers in the preface of the offending volume. But the committee grew increasingly impatient with the Agency's recalcitrance and eventually refused to approve twelve volumes that members considered misleading and incomplete. The impasse was broken when a hastily formed task force of senior State Department, CIA, and NSC officials voted to "acknowledge" fifteen of seventeen covert actions being considered for inclusion in the FRUS, thus clearing the way for the pertinent records to be selected and declassified.[25]

The conflict over the *Foreign Relations of the United States* series broke new ground in openness. The historians who led the charge on behalf of the *FRUS* law managed to turn countless individual scholars, with their widely differing but equally frustrating experiences of hitting the roadblock of excessive secrecy, into a coherent and effective force for change. And by granting private citizens access to secrecy's inner sanctum, Congress challenged the absolutism of the national-security establishment and sent a blast of fresh air through government vaults too long sealed against the public.

CONGRESS ACTS

In 1992, Congress took an even more forceful step in legislating openness when it passed the President John F. Kennedy Assassination Records Collection Act. The law was designed to end, once and for all, the deep-rooted public suspicion that the U.S. government was covering up critical information about the murder of the President in 1963. First of all, it directed simply that "all Government records concerning the assassination of President John F. Kennedy shall carry a presumption of immediate disclosure...." The law drastically limited exceptions to openness, requiring agencies to present "clear and convincing evidence" of harm if a record or fragment of a record was to continue to be withheld, and to balance that harm against the public's interest. The bill's third and most radical innovation was the creation of an independent review board of five private citizens to oversee the release of material. Unlike the *FRUS* advisory panel, the JFK board received unprecedented powers to review and declassify secret records (reversible only by the President), posing a direct challenge to the normal agency monopoly on information. The act represented a new frontier in the effort to balance legitimate secrets with the public's right to know. It was also a measure of how discredited the secrecy system had become. The text of the law contained an explicit rebuke of both the FOIA and the Reagan executive order, which, "as implemented by the executive branch, [have] prevented the timely public disclosure of records relating to the assassination of President John F. Kennedy."[26]

Beginning in April 1994, the JFK Assassination Records Review Board worked with dozens of federal, state, and local offices, foreign governments, and private citizens to secure the release of 60,000 documents—some 4 million pages—on the assassination and related matters. The board was not designed to solve the murder of President Kennedy. It was intended to produce the most comprehensive collection of records possible so that citizens could make their own informed decisions about what had happened on November 22, 1963. As documents began to be declassified, a steady stream of news articles based on the new information appeared. They covered not just the assassination but a huge range of topics, such as Vietnam and Cuba, Kennedy and the mob, covert activities in Central Europe and the Soviet Union, Pentagon plots against Castro, U.S. surveillance of Lee Harvey Oswald in the years before his arrest, and statements from the doctors who had performed the autopsy of the President. The review board sought to distinguish itself from the closed assassination investigations of the past by holding public hearings and inviting public comment on all its major decisions. Publicity about its work helped persuade private citizens who owned materials relating to the assassination to donate them to the collection, thus expanding it well beyond the scope of government holdings.

Predictably, the review board met fierce resistance from intelligence agencies. The board's final report, issued in September 1998, described its encounters with the absolutism of absolute secrecy: The CIA and the FBI both fought, for example, to suppress the names of all sources, dead or alive, and all methods, including such obvious and well-known methods as phone taps. Neither agency distinguished between the sensitivity of information gathered thirty years ago and that of information gathered three days ago. Both made the mistake of expecting the review board to accept without question their objections to disclosure. Without "clear and convincing evidence" of harm, however, the board decided it had no choice but to overrule the agencies and release the documents. The review board's power to do that—in the words of the final report, "to stand up to experts and their national security claims"[27]—made it an extraordinary experiment in openness.

There have been other congressional efforts to challenge the present-day secrecy system, primarily through the declassification of discrete collections of records where a compelling public interest existed, such as documents on Americans missing in Vietnam and on Nazi war crimes. Congress also launched a major review of U.S. secrecy and classification practices in 1995 with the establishment of the Commission on Protecting and Reducing Government Secrecy. Headed by Senator Daniel Patrick Moynihan, the so-called Moynihan Commission included the likes of Senator Jesse Helms of North Carolina; John Deutch, former Director of Central Intelligence; Ellen Hume, director of public television's "Democracy Project"; and Martin Faga, former chief of the National Reconnaissance Office. After two years of work, this unusual coalition produced a report that was full of plain truths about secrecy and democracy, though decidedly thin on meaningful proposals for change. The report contained an exhaustive history of U.S. secrecy and classification policies; condemned the contemporary secrecy system as expensive, anachronistic, and dangerous; and recommended reforms in the way the government creates and maintains secret information.

Despite the Moynihan Commission's progressive stance on openness, however, legislators have failed to institute any lasting structural reforms in the secrecy system during the first decade since the Cold War. And in the realm of intelligence information, Congress has missed numerous opportunities to limit the intelligence community's ability to cloak its activities in secrecy. That work was left to the executive branch.

Executive Tug-of-War

The nation's first post–Cold War President has taken some important first steps to challenge the secrecy system he inherited. Three months after his inauguration, President Clinton called for a review of the Reagan executive order on national-security information, asking that U.S. classification policy be "in line with the reality of the current, rather than the past, threat potential." When his Information Security Oversight Office invited public comment on the drafting of the new order, a surprisingly diverse crowd responded with calls for greater openness: press associations, freedom-of-information advocates, librarians decrying overclassification, defense contractors concerned about the mounting costs of secrecy, scholars seeking access to archives, and former intelligence officials demanding new ways to incorporate open-source information into intelligence-gathering. At one public forum, a panel of Air Force officers backed expanded declassification as a way to "reclaim our own Air Force history from a security file."[28]

Clinton also took the initiative early on to wave his presidential wand over several highly significant collections of secret documents. In November 1993, responding to calls from members of Congress and human-rights organizations, the administration released some 12,000 State Department, CIA, and Department of Defense documents on human rights in El Salvador. One year later, President Clinton signed an executive order opening almost 50 million pages of historical records, an enormous collection of previously secret material spanning the period from before the Second World War to Vietnam. The release represented the largest single declassification in U.S. history. It was followed by another milestone in openness, when Clinton required that satellite imagery from historical intelligence-reconnaissance missions be reviewed for release.

Sensing the change even before Clinton took office, agencies scrambled to renounce old secrecy habits, not wanting to be perceived as resistant to the new climate. It was beginning to dawn on the national-security bureaucracy that the era of the blank check was over. Agencies long submerged in the black waters of secrecy needed to surface quickly and become part of the growing public debate over changing missions and shrinking resources. In February 1992, Director of Central Intelligence Robert Gates went before the Oklahoma Press Association to announce the advent of "CIA openness" ("an oxymoron," he admitted), promising more news-media briefings, more academic conferences, and more documents. In September of the same year, the famously secret National Reconnaissance Office—the agency that for decades had dared not speak its name—quietly unveiled its existence in a one-page statement to the press.

For the most part, however, national-security agencies substituted spin for substance, working to alter public perception without assuming the burden of real action. Intelligence agencies, in particular, have rejected the trend toward openness. Born in secrecy, they are now not only struggling with an increasingly skeptical public that no longer believes their sweeping appeals to national security. They are also engaged in an internal battle between the old guard of covert operators, who resist exposure of any kind, and the reformers and political appointees, who are trying to take the first baby steps into the twenty-first century. Without effective congressional intervention, they are left to police themselves.

The CIA offers a case in point. Since its pledge to openness in 1992, the Agency has labored to overcome the credibility gap that fifty years of excessive secrecy have wrought. Robert Gates announced a shake-up of CIA declassification practices and revamped the historical-review program started in 1985 under Reagan's DCI, William Casey. The Agency's Center for Intelligence Studies began releasing a series of carefully selected, expensively bound collections of declassified documents on some of the CIA's oldest programs, such as the joint NSA-CIA Venona counterintelligence project aimed at Soviet espionage in the 1940s and '50s, the

Corona spy satellite of the 1960s, intelligence documents from the Truman era, and intelligence estimates on the Soviet Union. They were presented to the public with some fanfare during open conferences in Washington, D.C. In September 1993, Clinton's first CIA director, James Woolsey, testified before Congress that documents concerning eleven significant covert operations, all more than thirty years old, would be reviewed for release.

These are encouraging developments. But CIA openness goes only so far. Its glossy document collections are the result of an in-house initiative, and as such shine a decidedly favorable light on the Agency and its missions. The CIA voluntarily published its records on the intelligence success of the Cuban missile crisis, for example, but not on the covert failure of the Bay of Pigs. It publicized the daring Venona decryption program, but not its efforts to monitor domestic political dissent in the 1960s. Such selective historiography is the bane of scholars and archivists, the worst kind of "targeted declassification." The documents were not ordered released by Congress or the President on the basis of historical significance, as were the Kennedy assassination records and the files on Nazi war criminals, but were carefully chosen by the Agency itself and represent the Agency's own interests. The strategy has paid off. As Professor George Herring, one of the outside historians appointed by the CIA to serve on its Historical Review Panel, observed in 1997, "The Agency has done such a brilliant public relations snow job ... that in numerous conversations with people in and outside academia I was frequently told how the CIA was moving toward openness, a carefully nurtured myth that was not at all easy for me to dispel."[29] What the CIA's historical program has *not* yet achieved is something much less glamorous but infinitely more important: a plan establishing priorities and a timetable whereby the Agency can place complete collections of its most important office files in the hands of the public.

Efforts to reform the Agency from the outside quickly meet powerful institutional barriers. The barriers, first and foremost, are cultural—"rigid agency policies and procedures heavily biased toward denial of declassification," in the words of Robert Gates. They are also based in law. The CIA has been able to use its mandate to protect "intelligence sources and methods" as a shield against disclosure. Congress strengthened that shield in 1984 when it passed the CIA Information Act. The bill was an attempt by Congress to streamline the processing of FOIA requests to the Agency by exempting what was intended to be a small group of operational, security, and technical files from review. Its practical effect was to cordon off the Directorate of Operations altogether, making access to records about the CIA's covert activities almost impossible. Public access was dealt another blow a year later when a Supreme Court decision granted the Agency the right to hide even unclassified information as potentially damaging to "intelligence sources and methods." The result of *CIA v. Sims,* as Justice Thurgood Marshall wrote in a concurring opinion, was to cast an "irrebuttable presumption of secrecy over an expansive array of information in Agency files," giving the CIA "virtually unlimited discretion to label certain information 'secret.'"[30]

The extremism of the CIA's position is evident in its refusal to release its budget, a policy it has sworn by since the Central Intelligence Agency Act granted it the right to do so in 1949. The aggregate intelligence budget, which includes the CIA, the NSA, the NRO, and a handful of other agencies, has long been sought by critics of secret government spending. They charge that black budgets violate every citizen's constitutional right to a full account of the expenditure of public monies. Without such information, the people are deprived of an elemental insight into the workings of government and thus cannot debate them intelligently. Furthermore, accidental and deliberate disclosures of the budget over the years have made its continued secrecy especially puzzling. These arguments failed to sway the CIA, the courts, or Congress during the Cold War, but recently they have gained momentum. In 1996, President Clinton advocated publishing the figure "to promote openness in the Intelligence Community." The Director of Central Intelligence at the time, John Deutch, reaffirmed that position before the Senate, testifying that "disclosure ... will inform the public and will not, in itself, harm intelligence activities." Yet Congress has refused to order the release, leaving the CIA free to continue denying it as a matter of policy. It took a Freedom of Information Act lawsuit filed in 1997 to force the Agency to reveal that the total for that year was $26.6 billion. The new DCI, George Tenet, disclosed the budget the following year ($26.7 billion) and then, perversely, reversed himself by withholding the requested figure for 1999.[31]

And so it goes with most disclosures from the CIA: Unless they are voluntary, they have to be forced from the Agency by leaks, legal action, or scandal. Much of what we know about the CIA today entered the public realm against its will, from the bloodbath at the Bay of Pigs to the Iran-Contra debacle—epic policy failures that were themselves born of excessive secrecy. Even the CIA's own periodic gestures toward openness are thwarted from within. More than five years after James Woolsey vowed to release records on some of the Agency's historic covert missions, Tenet has managed to declassify only a fraction of the material on two of its operations, the Bay of Pigs and Guatemala. In July 1998, Tenet issued a press release declaring that the CIA did not have the resources to continue the declassification as promised, and would cease to focus on the covert operations.

Such a situation requires leadership. "It is my firmly held conviction that if there is to be real, significant change in the Intelligence Community, it will have to come from without, not from within," declared James Clapper, former director of the Defense Intelligence Agency, in 1995. "Because having recently been a part of it, I don't believe that the Community in and of itself is capable of fundamental reform."[32] So far, both Congress and the courts have shirked that responsibility.

President Clinton offers some hope for change. In 1995, two years after he launched his government-wide review of the country's secrecy policies, he signed Executive Order 12958, the first post–Cold War directive on the overall classification system of U.S. national-security information. The order destroyed one of the national-security establishment's most cherished beliefs—that secret documents must remain secret indefinitely—by requiring the "automatic declassification" of most historically valuable records that are twenty-five or more years old. Agencies have until the year 2000 to identify and set aside especially sensitive information, which would then be subject to the routine declassification process; everything else would be automatically disclosed. The order also required agencies to place a limit of ten years on most new secrets they create, thus rescinding the total discretion over the duration of secrecy that agencies enjoyed during the Reagan and Bush years. Finally, the executive order established the Interagency Security Classification Appeals Panel (ISCAP), a government review board with the power to reverse agency classification decisions. In its first two years of operations, ISCAP declassified in full or in part more than 80 percent of the withheld records that it reviewed. Combined with the early returns from automatic declassification—some 400 million pages released since the executive order was signed—the panel's actions represent an indictment of old secrecy practices and may be a harbinger of further openness.

For all of Clinton's achievements, however, he has not managed to bring the country up to the standards set by Jimmy Carter's executive order in 1977. The Carter order had a crucial element that Clinton's lacks: the public-interest balancing test that is so crucial to a real overhaul of the U.S. secrecy system. Nor does the Clinton order substantially narrow the exemptions that permit agencies to withhold entire categories of records from the public, making it unlikely that historically significant records about NSA intercepts, for example, will see the light of day anytime soon. Resistance to automatic declassification among national-security agencies is high. The FBI managed to win total immunity from automatic declassification for its estimated 7.8 billion pages of records, and the CIA has requested exemption of almost 65 percent of its files. Most recently, conservative members of Congress have asserted that automatic declassification places restricted nuclear data at risk for inadvertent disclosure; the objection has brought automatic disclosure to a halt for the time being.

Exhuming the Truth About Intelligence

If there is a lesson to be learned from the advances made during the past decade, it is that the future of secrecy reform rests in the hands of the people. Congress and the current President have, without a doubt, achieved a measure of change. Signposts pointing to a more rational system have been erected, and even the most closed agencies have made concessions to openness heretofore unthinkable. But popular pressure remains the single most effective weapon we have against the Frankenstein's monster of excessive secrecy.

The pressure has grown of late. The end of the Cold War has attracted new activists to the openness movement, as citizens galvanized by the desire to recover history from long-secret archives have joined the cause. Working in alliance with public-interest groups dedicated to freedom of information, this fresh wave of openness advocates has compelled some extraordinary disclosures and has helped push the limits of what may be released without damaging current U.S. security interests.

The campaign to obtain information on human-rights abuses in Latin America is a recent and fascinating example of the new directions that the search for openness is taking. The effort initially grew out of support for human-rights investigations under way in Central America. After the Cold War, nations throughout the region were wrestling with the violent legacies of past civil conflicts and repressive regimes. Decades of U.S. aid to those regimes made it likely that U.S. defense, diplomatic, and intelligence archives contained invaluable information for the Central American truth commissions. In 1993, the United Nations Truth Commission in El Salvador published a report on the brutal civil war in that country, drawing in part on declassified U.S. documents that had been released to American researchers under the Freedom of Information Act and provided to the commission. As a result, a coalition of U.S. public-interest and human-rights organizations began pressuring the Clinton administration to open its files on human-rights abuses and the perpetrators of those abuses. With newly democratic governments in Latin America willing to establish truth commissions and exhume the bodies of the victims of violence, the coalition argued, why couldn't the United States exhume the truth by releasing what it knows about the past?

In 1995, after revelations about CIA links to a murderous Guatemalan army officer, intense public pressure led to President Clinton's decision to ask his Intelligence Oversight Board to review U.S. intelligence operations in Guatemala. The resulting report, released in June 1996, documented U.S. covert support for torturers and murderers through its liaison relations with the Guatemalan military and intelligence services and through its network of paid assets inside the country. This was exactly the kind of information that U.S. intelligence agencies—in particular the CIA—have always insisted cannot be released due to the grave harm it would inflict on U.S. national security. Yet the Intelligence Oversight Board felt it was possible to disclose such details without damaging U.S. interests.

Requests for the release of human-rights information from the Cold War have now been taken up by other countries. In 1997, Spain submitted a formal petition to the United States for documents relating to its investigation of the Chilean dictator Augusto Pinochet. And in 1998, the Mexican congress requested U.S. information concerning the so-called Tlatelolco massacre, a student uprising of more than thirty years ago that ended in the deaths of hundreds of young Mexicans at the hands of Mexican security forces. Efforts to come to terms with the violence of the past have differed from nation to nation, but the search for truth has now come to include calls for historical records that might document past atrocities, identify perpetrators, and provide some measure of official acknowledgment to victims and their families. The demand for information is especially critical in societies that lack strong judicial systems and where historical accountability may be the most that people can expect.

If the United States seeks to grow as a democracy, its citizens must reject the obsession with secrecy that has dominated government operations for more than fifty years. We must demand that our Congress and our President broaden the incipient structural reforms that have been made to the secrecy and classification system to date, with the goal of finally dismantling the national-security state. The history of secrecy and the openness movement suggest a few key proposals for change that would help move our country toward the transparency and accountability required by a truly democratic society.

1. An appropriate balance must be struck between openness and secrecy in matters of national security. This requires three innovations: the presumption of openness, a public-interest balancing test, and the capacity for outside review. Though the Clinton executive order does incorporate a presumption of openness for most government records, it is inapplicable to intelligence information. That must change, either through an amendment to the order or through legislated limits on the intelligence community's ability to conceal information indiscriminately. (See below.) Likewise, a public-interest balancing test must be

an integral part of the classification and declassification process, through an amended executive order or through congressional action. Finally, the balancing test must be subject to judicial review; the courts must be required to consider the public's interest when deciding on challenges to agency denials under the Freedom of Information Act.
2. The classification practices of intelligence agencies require immediate reforms. First, the authority of the intelligence community to withhold information on the basis of "sources and methods" must be radically revised. Although intelligence agencies have the responsibility to protect legitimately sensitive information, the CIA's mere claim that disclosure would expose sources and methods cannot be sufficient. Congress should amend the National Security Act of 1947 so that the CIA is required to show harm under the classification standards set forth in the Clinton executive order.

Second, intelligence agencies must move to institute a historical-review process that results in the release of their oldest and most historically significant files to the National Archives. That means agreeing on document priorities and a schedule for declassification. The CIA's own Historical Review Panel has identified collections that the Agency could open, such as DCI office files for 1947–61, finished intelligence documents at least thirty-five years old, and all Directorate of Intelligence analytical studies on the Soviet Union.

Finally, intelligence-agency budgets must be declassified annually. The CIA's obstinacy on this point and the complicity of Congress in refusing to compel disclosure illustrate the extent to which the intelligence community remains apart from the normal practices of our democratic system. Congress must find the wherewithal to require that the CIA and all other intelligence agencies reveal their aggregate budgets.
3. FOIA offices must be adequately staffed and funded. If the Freedom of Information Act is to function as it was intended, agencies must request and allocate increased funding for the operations of their declassification units. Agency reviewers must not be permitted to use a lack of resources as an excuse for inordinate delays or improper denials in responding to FOIA requests.
4. There should be a clearly defined process regulating targeted declassification reviews. In the event of intense public interest in a specific collection of documents—such as those surrounding past human-rights abuses in Latin America—there should be a clear and expeditious process through which such documents can be reviewed for declassification. The process should include an outside review board, use precisely defined declassification standards, and be held to a specific timetable.

Taken together, these measures would help the public reassert its essential right to know about U.S. intelligence activities in particular and U.S. government operations in general. The American people deserve no less. What did the Cold War teach us, if not the fundamental instability of closed regimes?

That was the point made by a Clinton official at a recent conference on information and classification. The conference took place in November 1998, in the enormous, gleaming headquarters of the National Reconnaissance Office in Chantilly, Virginia. Roslyn Mazer, head of President Clinton's interagency panel on security classification, was speaking to an audience of several hundred U.S. intelligence officers. The Cold War ended, said Mazer, when closed regimes found themselves outpaced by open societies and collapsed from exhaustion: "This is the reason why our democracy endures, why we live under the oldest living constitutional democracy, and why we cannot export democracy like bananas to formerly closed societies. We prevailed over those societies because of our passion for openness, for trusting our citizens more than we empower our leaders. We celebrate our openness. In fact, it is unnecessary secrecy that is timid and cowardly. Openness is courageous.

"Be courageous."[33]

Notes

1. Senate Judiciary Committee, Subcommittee on Administrative Practice and Procedure, 88th Congress, 1st Session, *Freedom of Information* (October 28–31, 1963), p. 3.

2. Commission on Protecting and Reducing Secrecy, *Secrecy*, S. Doc. 105-2 (Washington, DC, 1997).

3. Information Security Oversight Office, *1997 Report to the President* (Washington, DC, August 31, 1998). For the full text of this document and additional information on developments in U.S. secrecy and classification policy, turn to the invaluable "Secrecy & Government Project" of the Washington-based Federation of American Scientists. Headed by Steven Aftergood, the project issues a bulletin and maintains a Web site at www.fas.org/sgp.

4. The law, passed in 1998, was part of the *Intelligence Authorization Act for FY 1999*, PL 105-272, Sections 701–702, "Whistleblower Protection for Intelligence Community Employees Reporting Urgent Concerns to Congress."

5. House Committee on Government Operations, Subcommittee on Foreign Operations and Government Information, 93rd Congress, 2nd Session, *Security Classification Reform* (July 11, July 25, and August 1, 1974), p. 356 (testimony given by William E. Colby).

6. And then there is the "10th FOIA exemption," as one reporter observed years ago: "We are not going to tell you." Senate Judiciary Committee, Subcommittee on the Constitution, 97th Congress, 1st Session, *Freedom of Information*

Act (1981), p. 495. Carl Stern quoted in testimony given by James Wieghart, American Society of Newspaper Editors.

7. CIA, *Task Force Report on Greater CIA Openness* (December 21, 1991). The report, and all declassified documents cited in this chapter, can be found in the holdings of the National Security Archive in Washington, DC.

8. In his recent book *Secrecy* (New Haven, CT: Yale University Press, 1998), pp. 84, 91–98, Senator Daniel Patrick Moynihan traces the origins of secrecy in the U.S. government to 1917, when the Espionage Act was being debated and signed into law. But although that law survived World War I, the greater part of the security system put into place was dismantled after the armistice. During the Second World War, on the other hand, secrecy was institutionalized; we continue to live in its shadow today.

9. House Government Operations Committee, 100th Congress, 2nd Session, *Presidential Directives and Records Accountability Act* (August 3, 1988). The quote from Speaker of the House Jim Wright appears in Jeffrey T. Richelson's essay on "National Security Policy and Presidential Directives," in the National Security Archive; *Presidential Directives on National Security: From Truman to Clinton* (Alexandria, VA: Chadwyck-Healey, 1994), p. 26.

10. Tim Weiner, *Blank Check: The Pentagon's Black Budget* (New York: Warner Books, 1990), pp. 120–121. See Weiner's chapter "Keeping Secrets" (pp. 111–142) for an elaboration of the secrecy system in general.

11. Jeffrey T. Richelson, "Out of the Black: The Disclosure and Declassification of the National Reconnaissance Office," *International Journal of Intelligence and Counterintelligence,* Vol. 11, No. 1 (Spring 1988): 1–2. Richelson quotes a top-secret Department of Defense directive from 1964 to illustrate the level of secrecy the NRO enjoyed: "[W]ith the single exception of this directive, no mention will be made of [the National Reconnaissance Office] in any document not controlled under the special security system(s) established to protect such information" (DOD Directive 5105.23, March 17, 1964).

12. Kathleen L. Enders, "National Security Benchmark: Truman, Executive Order 10290, and the Press," *Journalism Quarterly,* Vol. 67, No. 4 (Winter 1990): 1072.

13. NSC Intelligence Directive 12, *Avoidance of Publicity Concerning the Intelligence Activities of the U.S. Government* (January 6, 1950).

14. Matthew M. Aid, "Not So Anonymous: Parting the Veil of Secrecy About the National Security Agency" in Athan G. Theoharis (ed.), *A Culture of Secrecy: The Government Versus the People's Right to Know* (Lawrence: University Press of Kansas, 1988), pp. 70–71. The book as a whole was immensely helpful in writing this chapter, but in particular I would like to mention the following essays: James X. Dempsey, "The CIA and Secrecy"; Page Putnam Miller, "We Can't Yet Read Our Own Mail: Access to the Records of the Department of State"; and Anna Kasten Nelson, "The John F. Kennedy Assassination Records Review Board."

15. Herbert Feis, "Unpublic Public Papers," *New York Times,* April 21, 1968.

16. Senate Select Committee to Study Governmental Operations With Respect to Intelligence Activities, 94th Congress, 2nd Session, Report No. 94-755, Book I: *Foreign and Military Intelligence* (April 26, 1976), p. 12.

17. Executive Order 12065, "National Security Information" (June 28, 1978). Also see the President's "Statement on Issuing Executive Order 12065" in *Public Papers of the President: Jimmy Carter, 1978,* Vol. I. (Washington, DC: Government Printing Office, 1979), pp. 1193–1194.

18. Some of these initiatives were halted through the efforts of openness and civil-liberties groups. For a comprehensive review of the information policies of the Reagan administration, see the American Library Association's booklet "Less Access to Less Information by and about the U.S. Government: A 1981–1987 Chronology" (ALA Washington Office, February 1988). The ALA issues regular updates of its chronology, which can be obtained free.

19. "Iran-Contra Hearings: Summing Up a Summer's Work; Final Remarks by Leaders of the Panels: A Litany of Mistakes," *New York Times,* August 4, 1987, p. A6.

20. Blanche Wiesen Cook, "U.S. Foreign Relations History—Is There a Future at All?" *Perspectives,* November 1991, p. 12.

21. Kermit Roosevelt, *Countercoup: The Struggle for the Control of Iran* (New York: McGraw-Hill, 1979), p. 199.

22. Henderson to the Department of State, August 21, 1953. In *Foreign Relations of the United States, 1952–1954,* Vol. X, *Iran 1951–1954* (Washington, DC: Government Printing Office, 1989), p. 759.

23. Warren I. Cohen, "At the State Department, Historygate," *New York Times,* May 8, 1990, p. A29.

24. *Foreign Relations Authorization Act for FY 1992 and 1994,* PL 102-138, Section 404.

25. Page Putnam Miller, NCC *Washington Update,* Vol. 4, No. 41, October 15, 1998. For this or any other edition of this newsletter of the National Coordinating Committee for the Promotion of History—a key lobbying group during the *FRUS* controversy—see http://h-net.msu.edu/~ncc/.

26. *The President John F. Kennedy Assassination Records Collection Act of 1992,* Public Law 102-526. The law directed the President to choose the JFK board members after considering recommendations from four private professional associations: the American Historical Organization, the Organization of American Historians, the Society of American Archivists, and the American Bar Association.

27. *Final Report of the Assassination Records Review Board* (September 1998), p. 171.

28. Information Security Oversight Office, *1993 Report to the President* (Washington, DC, March 23, 1994), p. 14. The text of Presidential Review Directive 29 (April 26, 1993) is reproduced on page 2.

29. "My Years With the CIA," a speech given before the January 1997 meeting of the American Historical Association by George C. Herring, professor of history at the University of Kentucky in Lexington. The CIA's Historical Review Panel, composed of a group of leading U.S. historians, was established in 1984 to work with the Agency on identifying historically valuable records for declassification. Unlike the State Department's advisory committee, however, the CIA panel—in Herring's words—"had no chair, met at the whim of the Agency, exerted no real influence, and at times was used as window dressing."

30. *CIA v. Sims,* 471 U.S. 159, pp. 191–92 (1985).
31. Clinton and Deutch are quoted in *Aftergood v. CIA* (U.S. Dist. Ct., D.C. 1997), "Complaint for Declaratory and Injunctive Relief Under the Freedom of Information Act." The lawsuit was filed by the Center for National Security Studies on behalf of the Federation for American Scientists; the CNSS and the FAS countered Tenet's decision to deny the 1999 figure with a second lawsuit.
32. House Permanent Select Committee on Intelligence, *IC 21: The Intelligence Community in the 21st Century* (Washington, DC: Government Printing Office, April 9, 1996), p. 317.
33. Roslyn Mazer, head of the Interagency Security Classification Appeals Panel, in a speech before the 4th Annual Intelligence Community Information Management and Classification Conference, November 2, 1998.

ALFRED W. MCCOY

7 Mission Myopia
Narcotics as Fallout From the CIA's Covert Wars

IN AUGUST 1996, the *San Jose Mercury News* reported that CIA-supported Contra forces in Nicaragua had been involved in the cocaine traffic, sparking a major controversy and starting the first serious debate over the CIA's Cold War alliances with drug lords. In sum, the newspaper stated that during the 1980s a syndicate of Nicaraguan exiles had "sold thousands of pounds of cut-rate cocaine to Los Angeles street gangs and then used the lucre to buy arms for the Contras, the so-called freedom fighters."[1] In an editorial accompanying its exposé, the paper charged that "it's impossible to believe that the Central Intelligence Agency didn't know about the Contras' fund-raising activities in Los Angeles."[2]

At first, the story attracted little notice. But by mid-September, daily Internet "hits" at the *Mercury* had passed the million mark and African-American anger was rising. On talk radio, some black callers—going far beyond what the *Mercury News* had actually said—accused the CIA of willfully destroying their communities with crack. In Washington, D.C., the Congressional Black Caucus demanded an investigation.[3]

John Deutch, the CIA director, shot back that "the Agency neither participated in nor condoned drug trafficking by Contra forces." Unconvinced, the leader of the Black Caucus, Representative Maxine Waters of California, told an emotional meeting of 2,500 in the nation's capital that federal agencies "have responsibility" for the flow of illegal drugs. More than 2,000 African-Americans marched in Los Angeles, demanding that federal officials answer for their role in the crack epidemic. Representative Waters then wrote the U.S. Attorney General, charging that her city "may have been ... introduced to the horror of crack cocaine because certain U.S. Government ... operatives smuggled, transported and sold it to American citizens."[4]

As the controversy grew, the national press entered the debate on the side of the CIA. On October 4, the *Washington Post* published a front-page "investigation" denying that the rise of crack in Los Angeles was the work of just one syndicate, and charging that the *Mercury News*'s exposé merely "echoed decade-old allegations." Two weeks later, the

New York Times and the *Los Angeles Times* followed with their own investigations, attacking the *Mercury News*'s story and accusing that paper of fanning the flames of racial discord.⁵

In retrospect, this racially charged debate raises, perhaps for the first time, some important questions about the CIA and its covert alliances during the long decades of the Cold War. Did the Agency ever ally with drug traffickers? If so, did it protect these allies from investigation or prosecution? Did the CIA encourage drug smugglers to target African-American communities? And, above all, did these clandestine alliances contribute significantly to an expansion of the global drug trade?

Yet if our aim is to understand the underlying logic of the CIA's alliances with drug lords, the Contra case conceals as much as it reveals. During the months of debate over the *Mercury*'s report, most news commentators treated the Contras as an isolated case, ignoring the long history of the Agency's tactical alliances with Asian drug lords. Most important, the news media ignored the CIA's Afghanistan operation, whose reliance upon regional drug lords offers some revealing parallels with the Contra operation. Not only can this larger history shed light on the recent Contra-drug controversy, but it can open us to a broader assessment of the price that America paid for its use of CIA covert warfare to insure victory in the Cold War.

COLD WAR CONTEXT

Throughout the Cold War era, the CIA allied with gangsters and warlords, many of them drug dealers, to fight communism. This era has ended, but the list of Agency assets who used these alliances to traffic in narcotics became quite long, including Marseilles Corsicans, Laotian generals, Thai police, Nationalist Chinese irregulars, Afghan rebels, Pakistani intelligence, Haitian colonels, Mexican police units, and the Guatemalan military.⁶

Of all these alliances and their covert missions, the synergy between the CIA's operations and the drug traffic was most significant in Asia. In the late 1940s, communism's so-called iron curtain fell along an Asian opium zone that stretched for 5,000 miles from Turkey to Thailand, making these rugged highlands a key Cold War battleground. In Burma in the 1950s, Laos in the 1970s, and Afghanistan in the 1980s, the CIA relied upon highland warlords to mobilize tribal armies against communist forces. In each instance, local warlords exploited this alliance to become drug lords, expanding opium production and exporting growing supplies of heroin to international markets.

The Agency tolerated such trafficking and, when necessary, blocked investigations. Ruthless drug lords made effective anticommunist assets and heroin profits amplified their power, so CIA agents—operating alone half a world from home—did not tamper with the requisites of success in such delicate covert missions. When the CIA extended its paramilitary operations into Central America in the 1980s, using personnel and methods developed in Asia, its operations there bore a striking similarity in both conduct and consequences to earlier covert campaigns in Burma and Laos.

In Asia, distance insulated the Agency from the consequences of its complicity in the region's opium trade. But in Central America, geographical proximity and that region's particular role in the narcotics traffic would make the consequences of the CIA's compromises visible in ways that would spark public controversy.

These CIA operations in Asia and the Americas exhibit two central characteristics. By amplifying the political authority of the ethnic warlords, drug trafficking increased the CIA's operational effectiveness and improved its chances for short-term success. But over the longer term, these same operations produced a marked increase in local opium production and a major expansion in drug exports to Europe and the United States: a form of "fallout" with decidedly negative political and social consequences. If the CIA demonstrated a certain genius in its ability to manipulate local social conditions to mount its missions, it also exhibited something akin to "mission myopia" in its institutional insensitivity to the long-term consequences of these tactics. Indeed, surveying the steady increase in America's drug problem since the end of World War II, we can discern periodic increases in drug supply that coincide approximately with the CIA's covert operations in the drug-source countries.

Colonial Opium Trade

Though CIA operations often produced a marked expansion of drug trafficking, the Agency did not create the underlying social and economic conditions that made narcotics a significant factor in the political economy of covert war zones in Asia and Central America. Indeed, to place this recent controversy in its historical context, we need to understand how the CIA inserted itself into a particular socioeconomic setting, exploiting local conditions to facilitate its covert missions. Indeed, Agency operations often succeeded to the extent that they managed to manipulate existing local circumstances in ways that advanced its covert missions. We cannot assess these operations unless we understand both their local contexts and their lasting consequences. Assessing the CIA's complicity is best done by viewing its four decades of indirect involvement within the larger history of the global narcotics trade.

Over the past two centuries, opium has emerged as a major global commodity. The modern narcotics trade did not begin until the mid-eighteenth century, when the British East India Company, through its Calcutta office, took control of the export of Indian opium to China. By 1900, a nearly continuous band of opium cultivation stretched for 5,000 miles across the southern rim of Asia, from Turkey, through India, to Thailand, and across southern China.[7] By 1906, China alone was harvesting 35,000 tons of raw opium per year to supply 13.5 million opium addicts, equivalent to 27 percent of its adult male population.

The scandal of China's mass opium addiction prompted a global attempt at prohibition. Launched by Protestant denominations in the 1870s, the global anti-opium movement gradually built international support for an absolute prohibition of opium and coca products. In 1914, the U.S. Congress passed the Harrison Narcotics Act, for the first time making the use of heroin and cocaine illegal. During the 1920s, the League of Nations enacted regulations that gradually banned the sale of all narcotics and laid the foundations for the prohibition regime that would shape international narcotics traffic for the rest of the century.

Despite heavy opium consumption during the colonial era, Southeast Asia remained a minor producer. In 1940, the region that today harvests nearly 3,000 tons a year still produced only 15.5 tons of raw opium.[8] Why such a wide disparity between past and present? Since India supplied Southeast Asian monopolies with limitless, low-cost opium, governments had no reason to encourage local cultivation.

The sudden growth of Golden Triangle opium production in the 1950s appears, in retrospect, to have been a response to two stimuli: prohibition and protection. Between 1950 and 1961, under pressure from the United Nations, Southeast Asia's governments abolished legal opium sales, creating a sudden demand for illicit opiates in the region's cities. During this critical decade, an alliance of three intelligence services—Thai, American, and Nationalist Chinese—would play a catalytic role in promoting the production of raw opium supplies on the Shan Plateau of northern Burma.[9]

Burma

In the early 1950s, the CIA covert operations in northern Burma fostered political alliances that inadvertently linked the country's poppy fields with Southeast Asia's urban drug markets. After the collapse of the Nationalist Chinese government in 1949, some 14,000 troops fled across the border into northeastern Burma. To retaliate against communist China for its intervention in the Korean War, President Truman ordered the CIA to organize these Nationalist remnants for an invasion of southwestern

China. Despite a massive infusion of arms and supplies, the Nationalist Chinese irregulars mounted only three abortive incursions into China and were easily repulsed by local militia.

After their invasions of 1950 and 1951 failed, the Nationalist troops camped along the Burma borderlands for a decade, turning to opium to finance their operations. When the Burmese army evicted them in 1961, the Nationalist Chinese established new base camps just across the border in Thailand. From there they dominated the Shan state's opium trade until the early 1980s. By forcing local hill tribes to produce the drug through a mix of paramilitary coercion and market incentives, these Nationalist Chinese troops presided over a massive increase of poppy cultivation on the Shan Plateau—from eighteen tons in 1958 to between 400 and 600 tons by 1970.[10] In Thailand, General Phao Sriyanod, head of the national police and the CIA's local ally, provided the Kuomintang forces with logistics support and exported their opium shipments to markets in Southeast Asia and Hong Kong.

Kingdom of Laos

In Laos, as in Burma, distance insulated the CIA from the consequences of its complicity in the local drug trade. During France's war in Vietnam in the early 1950s, the French military had integrated opium trafficking with covert operations, forging a complex of covert-warfare alliances that the CIA would later inherit. After abolition of the opium monopoly in 1950, French military intelligence, the SDECE, imposed centralized controls over an illicit traffic that linked the Hmong poppy fields of Laos and Tonkin with opium dens operating in Saigon under the protection of the Binh Xuyen, a Vietnamese mafia allied with the French military.[11]

When America replaced the French in Vietnam after 1954, the CIA fell heir to these covert alliances. In Laos during the 1960s, the CIA battled communists with a secret army of 30,000 Hmong highlanders—a war that soon involved the Agency in that country's opium traffic. Although the CIA, unlike its French counterpart, did not profit directly from the trade, the combat strength of its tribal army was nonetheless integrated with the drug traffic.

At the start of its Laos operations in the early 1960s, a handful of CIA agents relied on tribal leaders to motivate their troops and Laotian generals to protect their clandestine cover. After the fighting in Vietnam spilled into Laos in 1965, the CIA-recruited Hmong army became a critical paramilitary asset. Between 1965 and 1970, the Hmong forces recovered downed U.S. pilots, battled the Pathet Laotian communist forces, and protected the radar that guided the bombing of North Vietnam. By 1971, according to a U.S. Air Force study, every Hmong family had lost at least one member and the tribe was faced with a demographic collapse.

This secret war was run by a handful of CIA agents—just one for every thousand Hmong guerrillas, a ratio that made the Agency dependent upon the tribal leaders, who could mobilize their people for a secret war that had become a bloody slaughter. To put it more precisely, by the late 1960s the success of the CIA's covert war in Laos rested upon one man: a Hmong officer in the Royal Lao Army named Vang Pao.

To prosecute a war that offered the ordinary Hmong soldier little more than rice and death, the CIA gave General Vang Pao control over all air transport into the tribal villages scattered across the mountaintops of northern Laos, including both the shipment of rice, the tribe's main subsistence commodity, into the villages, and the transport of opium, the tribe's only cash crop, out to markets in Vientiane and beyond. With a choke hold over the economic essentials of every Hmong household, General Vang Pao imposed a centralized command over this disparate tribe, quickly transforming himself from a minor officer into a powerful warlord who could extract boy soldiers for the slaughter in this hopeless war.

Since opium reinforced Vang Pao's authority, pragmatism dictated that the CIA should tolerate the traffic. This compromise might have been understandable at the outset, in 1960 or even 1965, but it became less comprehensible when the Agency's Laotian allies began producing heroin for U.S. combat forces fighting in South Vietnam. In 1968–69, CIA clients opened a cluster of seven heroin laboratories in the Golden Triangle—the region where Burma, Thailand, and Laos converge. When Hmong officers loaded opium on the CIA's own Air America, and the commander of the Laotian army, General Ouane Rattikone, opened a major heroin laboratory to supply U.S. troops in Vietnam, the Agency was silent. In a secret internal report, compiled in 1972, the CIA's Inspector General explained the reasons for the Agency's inaction: "The past involvement of many of these officers in drugs is well known, yet their goodwill ... considerably facilitates the military activities of Agency-supported irregulars."[12]

Instead of trying to restrain drug trafficking by its Laotian assets, the CIA engaged in concealment and cover-up. When I went to Laos in 1971 to do research for a book, the Laotian army commander graciously opened his opium accounts for examination but the U.S. embassy stonewalled, insisting that the commander had never been involved in the drug trade. When we were in a remote village investigating Hmong opium shipments on the Agency's Air America helicopters, CIA mercenaries ambushed my research team. Several days later, a CIA operative threatened to murder my Laotian interpreter unless I abandoned my investigation. After the book was published, CIA agents in Laos pressed my sources to recant and convinced the House Foreign Affairs Committee that my allegations were baseless.

Simultaneously, however, the CIA's Inspector General conducted a secret internal investigation that confirmed my findings. "The war has clearly been our overriding priority in Southeast Asia and all other issues have taken second place," he said in defense of the Agency's inaction on drugs. "It would be foolish to deny this, and we see no reason to do so."[13]

Heroin from the Laotian laboratories was smuggled into South Vietnam, where by 1971, according to a White House survey, 34 percent of U.S. troops were addicted.[14] If we accept this figure, there were some 80,000 American heroin addicts in South Vietnam, far more than the 68,000 in the United States, all supplied by America's Southeast Asian allies.

The involvement of our Southeast Asian allies in the heroin trade, and the CIA's complicity, should have been cause for protest and a serious congressional investigation. But Asia was too remote and its connection to the U.S. drug market too distant for reports of CIA involvement in the traffic to inspire a serious attempt at reform.

Nixon's War on Drugs

As American troops were withdrawing from Vietnam in 1972, President Nixon inadvertently created a new market for Southeast Asian heroin by declaring war on drugs in the Mediterranean. Acting on reports that Turkey's poppy fields and Marseilles's heroin laboratories supplied 80 percent of America's demand, President Nixon pressed Turkey and France to eliminate the traffic.

By 1975, Turkey, with $30 million in U.S. aid, had eradicated all opium production. Simultaneously, the street price in New York tripled and purity dropped by half: strong indicators of a serious shortage. Nixon had achieved a major victory in America's first drug war. Nixon's success in Turkey, however, increased global drug demand by unleashing market forces that would ultimately expand both production and consumption of illicit narcotics. Responding to the market stimulus of Turkish eradication, the Chinese syndicates of Southeast Asia began exporting their surplus heroin and had soon captured a quarter of America's illicit market.

Although Nixon's drug war stimulated the global market, its effect on U.S. domestic demand was more ambiguous. By the late 1970s, every indicator pointed to a marked decline in U.S. heroin supply. In New York City, the estimated number of heroin addicts dropped from 500,000 to 200,000, and heroin purity declined to a low of 3 percent. As total U.S. heroin imports decreased from seven to four tons, the number of addicts

Mission Myopia 125

nationwide fell from 750,000 in 1977 to some 400,000 two years later.[15] Why the decline? We cannot, in fact, answer with any certainty. Perhaps Nixon's drug war actually succeeded in slowing the flow of drugs into the United States. Or perhaps President Jimmy Carter's ban on major CIA covert operations from 1976 to 1978 removed the protection that drug lords seem to require.

NARCOTICS IN THE 1980S

Whatever the cause of the decline in the 1970s, prohibition and protection again played a role in the revival of the U.S. drug problem in the 1980s. In 1979, the Soviets invaded Afghanistan and the Sandinistas seized Nicaragua, prompting two major CIA operations with some revealing similarities. Significantly, these covert wars seemed to play a catalytic role in stimulating narcotics production in drug-source regions that became major suppliers for the United States during the 1980s.

CIA covert operations in Afghanistan contributed to the transformation of Central Asia from a self-contained opium market into a major international supplier of heroin. In this decade of Cold War confrontation with the Soviet Union, CIA support facilitated the political protection and logistics linkages that connected Afghanistan's poppy fields to heroin markets in Europe and America. Even after the superpowers disengaged in the early 1990s, Afghan traffickers, like those in Burma, used logistics and linkages forged in a decade of covert warfare to maintain and even expand their role in the global heroin traffic.

Unlike in most of Asia, European colonialism had little effect upon the tribal highlands of Afghanistan. The combination of rugged terrain, a population of well-armed "martial tribes," and a deep Islamic faith made Central Asia impervious to conquest.[16] With such weak control, Britain minimized its presence in India's Northwest Frontier region to avoid provoking the tribes. Instead of encouraging opium production for revenue, as they did elsewhere in India, the British saw the drug as a destabilizing factor in the Northwest Frontier and worked toward a gradual abolition of cultivation by 1901.[17] At the end of the colonial era in 1947, Afghanistan and Pakistan were minor opium producers with small addict populations.

The independent kingdom of Iran, in contrast, had become one of the world's major opium producers and consumers. Although the government banned opium smoking in 1910 and imposed prohibitive taxes in 1928, neither measure found much support from a population that accepted recreational drug use without reservation.[18]

When the U.S. Bureau of Narcotics opened an office in Tehran in 1949, Garland Williams, the agent in charge, discovered the world's most elaborate drug culture. Drawn from all social classes, Iran's addict population numbered 1.3 million, or one opium user for every nine adults—a rate exceeded only by that of China. In a tour of the capital, Williams found evidence of 500 public opium dens with a capacity for 25,000 smokers.[19] Seeking to eradicate this endemic addiction, the Shah imposed a complete ban on the production and consumption of opium in 1955. In 1969, after fourteen years of an imperfect ban plagued by high costs and corruption, the Shah announced that Iran would resume opium cultivation for sale to registered addicts, a decision that aroused opposition from the United Nations and the United States.[20]

The Shah's decision simply legitimated a portion of Central Asia's self-contained opium trade. By 1972, Iran's addict population had grown to some 400,000, of whom 105,000 were registered opium smokers. The authorized addicts consumed the country's entire opium crop of 217 tons, while illegal smokers used some 195 tons of illicit opium smuggled from Turkey, Afghanistan, and Pakistan.[21]

Though opium consumption was high, Iran's role in the international traffic was limited. The U.S. Ambassador, Richard Helms, who had recently retired as director of the CIA, reported in 1974 that Iran's "considerable population of unregistered users ... acts as a sponge for opiates produced elsewhere in the Middle East, thereby diverting supplies that might otherwise find their way as heroin to the United States."[22] In June 1975, in one of his last reports, Ambassador Helms argued that the region's self-contained opium trade was still no threat to the West. On the whole, Helms found that Iran's system of legal sales to registered opium addicts "works fairly well." Only a small group was processing heroin locally, and there was "no evidence" that either opium or heroin was being exported from Iran.[23] All available evidence thus indicates that as of the late 1970s, the historic patterns of Asia's opium trade had not changed and the region's drug traffic was still self-contained.

Within only five years, however, Central Asia's opium trade would experience a transformation. By the mid-1980s, Afghanistan had become the world's second-largest opium producer, the Pakistan-Afghanistan border was the leading source of heroin for Europe and America, and mass heroin addiction was spreading rapidly in Pakistan. Although the CIA did not play the same paramount role that it had in the transformation of Burma's drug trade thirty years before, its covert warfare still served as a catalyst in the emergence of Central Asia as a leading source of heroin for the world market.

In 1978 and '79, Central Asia became the focus of a major crisis in U.S. foreign policy. In February 1979, Tehran mobs toppled the Shah's dictatorship, destroying a conservative regime that had long served as

America's proxy military presence in the strategic Persian Gulf. Compounding the crisis, Soviet troops invaded Afghanistan in December, occupying the capital, Kabul, and installing a pliable Afghan communist as President. As Soviet troops occupied Afghanistan, President Carter used diplomatic and covert resources to mobilize military aid for the mujahideen guerrillas. Massive arms shipments began within weeks: handheld missiles and antitank weapons from China, Kalashnikov assault rifles from Egypt, munitions from Saudi Arabia, and a variety of U.S. weapons from the CIA.[24]

With Pakistan serving as the frontline state, the military regime of President Mohammad Zia ul-Haq assumed a dominant role in supplying the Afghan resistance, with the CIA working through Pakistan's Inter-Service Intelligence (ISI).[25] At the time General Zia took power in 1977, the ISI was a minor military intelligence unit with an annual budget of no more than several million dollars. With the "advice and assistance of the CIA," Zia soon built it into a powerful covert-operations unit and made it the strong arm of his martial-law regime.[26] Through this alliance, U.S. covert aid to the Afghan guerrillas began in early 1979, six months before the full Soviet invasion, when a CIA special envoy first met with Afghan resistance leaders, all carefully selected by the ISI, at Peshawar in Pakistan's Northwest Frontier Province.[27] Instead of presenting a broad coalition of resistance leaders, the ISI offered the CIA's envoy an alliance with its own Afghan client, Gulbuddin Hekmatyar, leader of the small Heb-i Islami guerrilla group. Surprisingly, the CIA accepted the offer, and over the next decade it gave more than half its covert aid to Hekmatyar's extremist Muslim faction. It was, as the U.S. Congress and press would learn a decade later, a dismal decision. Unlike the later resistance leaders, who commanded strong popular followings inside Afghanistan, Hekmatyar's guerrilla force was a creature of the Pakistan military. After the CIA built his Heb-i Islami into the largest Afghan guerrilla force, Hekmatyar would prove himself brutal, incompetent, and corrupt.[28] Not only did he command the largest guerrilla army, Hekmatyar would use it—with the full support of the ISI and the tacit tolerance of the CIA—to become Afghanistan's leading drug lord.

Once again, the CIA was mounting a major covert war in the remote highlands of the Asian opium zone—this time Pakistan's Northwest Frontier Province, a tribal territory adjacent to Afghanistan. "In the tribal area," the U.S. State Department reported in 1986, "there is no police force. There are no courts. There is no taxation. No weapon is illegal.... Hashish and opium are often on display."[29] With some 300 trails and passes leading to Afghanistan and, in the CIA's words, "the largest expanse of territory in the world without any government presence," this remote region was an ideal base for the Agency's secret war against Soviet expansion.[30] As in Burma and Laos, the CIA decided to conduct its war through a single local commander, making its success synonymous with his power. Under such circumstances, geographical and political, the CIA had little leverage when its tribal warlord decided to exploit the clandestine operation to become a drug lord.

Thus, when the CIA envoy agreed to provide some arms to Hekmatyar's guerrillas in May 1979, it was, although nobody knew it at the time, a momentous decision. After the Soviet army invaded Afghanistan in December, U.S. covert aid took a quantum leap—but along lines set six months before. In effect, the bargain struck at Peshawar in May 1979, which was confirmed by executive agreement between Presidents Zia and Reagan two years later, bound the White House and its covert-action arm to subcontract its secret war in Afghanistan to Pakistan's military. With generous U.S. aid, Pakistan opened its border to 3 million Afghan refugees and allowed the CIA to conduct its secret war without restraint. Along the border, American operatives ran training camps for the mujahideen, and in the capital, Islamabad, the CIA maintained one of its largest foreign stations to direct the covert war. Aside from the $3 billion in conventional U.S. aid, the Pakistan military gained control over delivery of the $2 billion in covert aid that the CIA shipped to the Afghan guerrillas during the ten-year war. For President Zia's loyalists within the military, these contracts were a source of vast wealth.[31]

At an operational level, Zia's military loyalists controlled the delivery of the CIA's covert arms shipments when they arrived in Pakistan. Once the arms landed at the port of Karachi in the south, the Pakistani army's National Logistics Cell, acting under orders from the ISI, trucked them north to military cantonments around Peshawar and from there to the Afghan guerrilla camps along the border. The governor of the critical Northwest Frontier Province was Lieutenant General Fazle Haq, President Zia's closest confidant, the de facto overlord of the mujahideen, and the distributor of these arms.[32] Even as the ranks of the resistance swelled after 1981, the ISI still insisted that Hekmatyar be given the bulk of CIA arms shipments. From the outset, CIA case officers were aware that the ISI's distribution system "was creating a form of warlordism within the resistance command."[33]

Within two years, the covert supply apparatus set up to ship arms into Afghanistan had been inverted to serve—and mask—a massive drug operation that moved opium from Afghanistan through heroin laboratories in Pakistan's Northwest Frontier and into international markets. As a network of heroin laboratories opened along the Afghan-Pakistani border to service Western markets in 1979–80, Pakistan's opium production soared to 800 tons, compared with a 1971 harvest of some ninety tons. By 1980–81, Pakistani-Afghan heroin dominated the European market and supplied 60 percent of U.S. illicit demand.[34]

As the mujahideen used their new CIA munitions to take control of prime agricultural areas inside Afghanistan during the early 1980s, the guerrillas pressed their peasant supporters to grow poppies, thereby doubling the country's opium harvest to 575 tons by 1983.[35] During most of the war, the local commander, Mullah Nasim Akhundzada, controlled the irrigated lands of the lush Helmand valley, once the breadbasket of Afghanistan; he decreed that half of all peasant holdings would be planted to opium. In a month-long trip through the Helmand valley in early 1986, Arthur Bonner, a *New York Times* correspondent, found extensive poppy fields in every village and town. Mohammed Rasul, older brother of the mullah, explained to Bonner, "We must grow and sell opium to fight our holy war against the Russian non believers."[36] Visitors to Helmand during this period spoke "in awestruck tones of the beauty of the poppies which stretch mile after mile."[37] Farther south, in the Koh-i-Soltan district of Pakistan's Baluchistan Province, Hekmatyar himself controlled six heroin refineries that processed the large opium harvest from the Helmand valley.[38]

Once the mujahideen brought the opium across the border, they sold it to Pakistani heroin refiners who operated under the protection of General Fazle Haq. By 1988 there were 100 to 200 heroin refineries in the province's Khyber district alone.[39] Trucks from the Pakistani army's National Logistic Cell, arriving with CIA arms from Karachi, often returned loaded with heroin—protected from police search by ISI papers. "The drug is carried in N.L.C. trucks, which come sealed from the N.W.F.P. and are never checked by the police," reported *The Herald* of Pakistan in a September 1985 issue. Writing in *The Nation* three years later, Lawrence Lifschultz cited numerous police sources to charge that General Fazle Haq was the primary protector of the thriving heroin industry in the Northwest Frontier Province, noting that he "had been implicated in narcotics reports reaching Interpol" as early as 1982.[40]

The heroin boom was so large and uncontrolled that drug abuse swept Pakistan itself in the early 1980s, creating the world's largest addict population. In the late 1970s, Pakistan had had almost no heroin abuse at all. But local smugglers, unrestrained by police, began shipping heroin to the country's own cities and towns—raising addiction from 5,000 users in 1980 to 1.3 million addicts in 1985.[41]

At the outset of this operation in 1983, Washington's implicit choice to sacrifice the drug war to fight the Cold War was articulated clearly in congressional hearings over the Reagan administration's request for $583 million in aid for Pakistan. Since the Rodino amendment required suspension of aid to any country "not taking adequate steps to control production of shipment of illicit narcotics," the administration had to convince Congress that General Zia was our ally in the drug war. Steven Solarz, chairman of a House Foreign Affairs subcommittee, put the problem succinctly in a question to an administration witness: "Considering the fact that 60 to 70 percent of the heroin on the streets of our country today reportedly comes from Pakistan, have you considered the political implications of my voting several billion dollars for Pakistan when my constituents are plagued by drug addiction and the Pakistanis do not appear on their way to solving the problem?"

In his testimony, Dominick DiCarlo, Assistant Secretary of State for Narcotics, stated that Pakistan had become "the prime opium source for U.S. heroin" and that some fifty laboratories in the Northwest Frontier Province now accounted for more than half the heroin used by America's 500,000 addicts. Nonetheless, DiCarlo assured the committee that the administration was "very serious about our approach to narcotics," and said that the Pakistani problem had "been taken up at the highest level." During visits to the United States, for example, "President Zia and NWFP governor Fazle Haq restated Pakistan's commitment to its opium poppy ban." In questioning, DiCarlo described in some detail the visit of Governor Haq to Paris in 1980 to seek assistance for opium eradication. He concluded that Pakistan would pass the criteria for certification "with flying colors," and warned that pressing our ally further on the issue would "be disastrous." Similarly, a senior official of the U.S. Agency for International Development, Charles Greenleaf, hailed President Zia for his "noteworthy progress in reducing opium production." Articulating the administration's real priority, Representative Charles Wilson argued that Pakistan needed our support as a frontline state housing nearly 3 million Afghan refugees, adding that if we gave the Afghans "a tenth of what the Russians provided to the Vietnamese, the Russians would really have their hands full."[42]

Washington's priority was reflected in a weak anti-narcotics effort by the Drug Enforcement Administration in Islamabad. With heroin processing booming beyond control, it did not take a great deal of police work to grasp the political realities of Pakistan's drug trade. Indeed, as Pakistani heroin flooded Europe, a single Norwegian detective broke a drug case in 1984 that led directly to President Zia's household. After a Pakistani smuggler was arrested at the Oslo airport with 3.5 kilograms of heroin and traded information for a reduced sentence, Oyvind Olsen, an Oslo detective, flew to Islamabad and soon found ample evidence in the country's wide-open drug trade. In September 1985, Norway's public prosecutor filed formal charges against three Pakistani heroin merchants, and Pakistan's Federal Investigation Agency (FIA) arrested Hamid

Hasnain, vice president of the government's Habib Bank, who was found with President Zia's personal banking records in his possession. On the night of the arrest, the President's wife called senior FIA agents from Egypt to demand Hasnain's release. In the end, however, the President's banker and friend was convicted and sentenced to a long prison term.[43]

The DEA unit in Islamabad did much less with seventeen agents, sixteen additional staff members, and a budget of $19.6 million between 1985 and 1988. Though the DEA compiled detailed reports identifying "forty significant narcotics syndicates in Pakistan," it did not mount any serious investigations or participate in any arrests while the CIA was operating in the Northwest Frontier Province. Indeed, in 1988, the U.S. General Accounting Office reported that "not a single significant international Pakistani trafficker is known to have been imprisoned prior to 1984," and that those jailed after that date "were quietly released after serving a few months."[44]

The blatant official corruption continued until August 1988, when President Zia's death in an air crash allowed an eventual restoration of civilian rule. Typical of the misinformation that had blocked any U.S. action against Pakistan's heroin trade, the State Department's semiannual narcotics review in September of that year called Zia "a strong supporter of anti-narcotics activities in Pakistan" and speculated that his death might slow the fight against drugs.[45]

After ten years of unchecked growth under Zia, Pakistan's heroin trade was now too well entrenched in the country's politics and economics for simple law enforcement or ready political reform. Conservative economists estimated the total annual earnings from Pakistan's heroin trade at $8 billion to $10 billion, far larger than Pakistan's government budget and equal to a quarter of its entire gross domestic product, giving drug syndicates ample resources to corrupt the country's politics.[46] Moreover, the heavily armed tribal populations of the Northwest Frontier Province were determined to defend their opium harvests.[47]

Even if Pakistan were to eradicate its entire yearly harvest of 130 tons of opium, Afghanistan's harvest, the world's second-largest at 800 tons in 1989 and rising rapidly, would have easily been able to expand production to supply Pakistan's network of heroin laboratories.[48] The Soviet withdrawal in February 1989 and a slackening in CIA support produced a scramble among rival mujahideen commanders for Afghanistan's prime opium land, most notably in the fertile Helmand valley. During most of the war, Mullah Nasim Akhundzada, the local commander known as the "King of Heroin" and later the deputy defense minister in the U.S.-backed Afghan interim government, controlled most of the 250 tons of opium grown in Helmand Province.[49]

While Mullah Nasim ruled the opium fields of Helmand, Hekmatyar held the heroin complex at Koh-i-Soltan at the southern end of Helmand, just across the border inside Pakistan. Beginning in 1988, Hekmatyar's local commander challenged Mullah Nasim's rule over the Helmand opium harvest. Once the snows melted in the spring of 1989, the war revived, now focused on a bridge that linked Helmand to Pakistan's heroin refineries. In the savage fighting, both sides absorbed heavy casualties, crippling their later offensive against the Soviet-backed regime.[50] Though Mullah Nasim won that opium war and inflicted heavy losses on Hekmatyar's forces, within a year he, too, was dead, probably from a drug assassin's bullet.[51]

After a decade of silence, the international press finally began to reveal the involvement of the Afghan resistance and Pakistani military in the region's heroin trade. In May 1990, the *Washington Post* published a page-one article charging that the United States had failed to take action against Pakistan's heroin dealers "because of its desire not to offend a strategic ally, Pakistan's military establishment." The paper reported that "Hekmatyar commanders close to ISI run laboratories in southwestern Pakistan," and that "ISI cooperates in heroin operations." Significantly, the *Post* claimed that U.S. officials had refused for many years to investigate charges of heroin-dealing by Hekmatyar and the ISI largely "because U.S. narcotics policy in Afghanistan has been subordinated to the war against Soviet influence there."[52] Just two months later, in a July issue, *Time* magazine claimed that the United States is "embarrassed by the widely bruited connections between the drug trade and ... Gulbuddin Hekmatyar's Hezb-i Islami," since "the U.S. is prosecuting an uncompromising war against drugs elsewhere in the world."[53]

More recently, former CIA operatives have admitted that this operation led to an expansion of the Pakistan-Afghanistan heroin trade. In 1995, the former CIA director for this Afghan operation, Charles Cogan, spoke frankly about the Agency's choices. "Our main mission was to do as much damage as possible to the Soviets. We didn't really have the resources or the time to devote to an investigation of the drug trade," he told Australian television. "I don't think that we need to apologize for this. Every situation has its fallout.... There was fallout in term of drugs, yes. But the main objective was accomplished. The Soviets left Afghanistan."[54]

Again, distance insulated the CIA from the political fallout of its compromises. Once the heroin left Pakistan, the Sicilian Mafia exported it to the United States, where mafiosi distributed the drugs through pizza parlors and local gangs sold it on the street. Most Americans could not do the political calculus needed to equate Afghan drug lords and Mafia pizza parlors with the heroin in their cities.

THE CONTRA OPERATION

In Central America, however, simple proximity made the fallout from the CIA's alliance with the anticommunist Contra guerrillas explosive. Unlike the CIA's Asian warlords, Nicaragua's Contras did not produce drugs themselves, and had to make money by smuggling cocaine through Central America into the United States.

This proximity brought these operations to the attention of Congress and the press. In 1985, the Associated Press carried a story by Brian Barger and Robert Perry stating that "Nicaraguan rebels operating in northern Costa Rica have engaged in cocaine trafficking, in part to help finance their war against Nicaragua's leftist government."[55] Several years later, Senator John Kerry's subcommittee conducted an investigation of these alleged Contra-cocaine links. His investigators established that four Contra-connected corporations hired by the U.S. State Department to fly "humanitarian relief" goods to Central America were also involved in drug smuggling. The pilots themselves gave eyewitness testimony that they had seen cocaine loaded on their aircraft for the return flight to the United States.[56]

The DEA operative assigned to Honduras, Thomas Zepeda, testified that his office had been closed in June 1983, since it was generating intelligence that the senior military officers were involved in cocaine smuggling—information that threatened the CIA's relationship with the Honduran military in this key frontline state in the war against Nicaragua's Sandinista regime. In 1981, the DEA, responding to the growing importance of the country as a transshipment point for Colombian cocaine, had opened a new office in the Honduran capital, Tegucigalpa. During his two years as the DEA's chief agent in Honduras, Zepeda found that the country's ruling military officers were involved in the cocaine traffic and did everything possible to slow his inquiries. "It was difficult to conduct an investigation and expect the Honduran authorities to assist in arrests when it was them we were trying to investigate," Zepeda told the U.S. Senate. In June 1983, the DEA closed its Honduran office without consulting Zepeda and transferred him to Guatemala, "where he continued to spend 70% of his time dealing with the Honduran drug problem." In his testimony before Kerry's committee, Zepeda said that he would have argued, if asked, that the DEA's Honduran office should have remained open, since he had "generated a substantial amount of useful intelligence."[57] Only four years later, Honduras had become such an important drug-transshipment point that the DEA was forced to reopen its Tegucigalpa office.[58] In effect, Agency operations had once again created a zone of protection, closed to investigators from outside agencies. Read closely, the Kerry committee's report established a pattern of CIA complicity in Central America strikingly similar to the situation in Laos ten years before: tolerance for drug dealing by the Agency's local assets and concealment of this criminal activity to protect its larger covert operation.

In its 1996 investigative report, the *San Jose Mercury News* tried to establish the next, critical link in this cocaine chain by showing a direct connection between the Contra war and the distribution of drugs in the United States. According to Gary Webb, a *Mercury News* reporter, this "dark alliance" began in the early 1980s, when the Contra revolt against Nicaragua's leftist Sandinista government was failing for want of funds. In 1981, the CIA hired Enrique Bermudez, a former Nicaraguan army colonel, to organize what became the main Contra guerrilla army, the FDN. Bermudez then turned to two Nicaraguan exiles in the United States to supplement meager CIA support with drug profits.

In California, Danilo Blandon, the former director of Nicaragua's farm marketing program, used his formidable business skills to open a new drug-distribution network for the Contras. Sensing the potential of the Los Angeles ghetto, Blandon allied with the then-neophyte, now famous, African-American dealer "Freeway Rick" Ross to convert tons of cocaine into low-cost crack and exploit what was still an untapped market among the city's poor blacks. When they started dealing in Los Angeles's South Central district, the *Mercury News* reported, crack "was virtually nonexistent." With limitless supplies of cheap cocaine from Central America, Ross soon built a booming drug business that spread up the California coast and across the Midwest.

During its decade of operation, this crack-distribution network, according to the *Mercury News*, enjoyed de facto immunity from prosecution. Whenever the DEA, the Customs Bureau, the Los Angeles County Sheriff, or the U.S. Congress tried to investigate, the CIA and the Justice Department denied information on grounds of "national security." In 1986, Los Angeles sheriff's officers raided what their warrant called Blandon's "sophisticated cocaine smuggling and distribution operation," but found every location wiped clean of evidence. The police were convinced that their investigation "had been compromised by the CIA." By the late 1980s, the operation had lost its Contra connection and both dealers were soon arrested on drug charges. Freeway Rick started serving a ten-year sentence, but the Justice Department intervened to free the Contra-connected Danilo Blandon.[59]

The CIA's relations with Asian opium lords had long been lost in far-off mountains. But Central America's proximity allowed Representative Maxine Waters, the Black Congressional Caucus leader from Los Angeles, to discover police documents implicating the CIA in the protection of cocaine dealers connected to their Contra operation.[60] Since the Agency's assets were, in this case, involved in the distribution of drugs at street level in America, these charges aroused a storm of controversy even a decade later.

Despite the CIA's denials and a skeptical reaction from the mainstream press, this controversy, in its waning hours, produced important insights into the dynamics of the Agency's alliances with drug lords. In response to the outrage prompted by the *Mercury News* series, Director of Central Intelligence Deutch had ordered Frederick P. Hitz, the Agency's Inspector General, to open an investigation. For nearly eighteen months, Hitz's seventeen investigators reviewed 250,000 pages of documents and conducted 365 interviews to produce a two-part report. In presenting volume one of his report to Congress in March 1998, the Inspector General found "absolutely no evidence to indicate that the CIA as an organization or its employees were involved in any conspiracy to bring drugs into the United States."[61]

Though the Agency had already leaked this story to several reporters, and some papers had reported that the *Mercury News* series was officially discredited, the inspector closed his testimony with an important admission that received surprisingly little coverage.[62] "Let me be frank," Hitz said. "There are instances where CIA did not, in an expeditious or consistent fashion, cut off relationships with individuals supporting the contra program who were alleged to have engaged in drug trafficking or take action to resolve the allegations." He also revealed that the CIA had, at the start of the Contra operation in 1982, concluded a "Memorandum of Understanding" with U.S. Attorney General William French Smith, exempting the Agency from reporting drug trafficking by its nonemployees, meaning "assets, pilots who ferried supplies to the contras, as well as contra officials and others." Even though this agreement was modified four years later to require some reporting, the CIA continued to operate with these individuals, and investigations into their drug dealing were not done "as expeditiously as they should have been." Remarkably, the waiver remained in effect until 1995.[63] Several months later, the *New York Times* reported that the inspector's still-classified volume two found that the CIA had worked with "about two dozen Nicaraguan rebels" who were known to be "trafficking in drugs."[64]

In October 1998, the CIA finally posted a heavily censored version of the Hitz report's volume two on its Web site, revealing that the Agency had been aware of the Contras' involvement in drug dealing from the very outset and had done nothing to stop it. "In September 1981," said the *New York Times,* paraphrasing the Hitz report, "as a small group of rebels was being formed from former soldiers in the National Guard of the deposed

Nicaraguan dictator, Anastasio Somoza Debayle, a CIA informant reported that the leadership of the fledgling group had decided to smuggle drugs to the United States to support its operation."[65] Though Congress later gave the CIA direct orders to report any drug dealing by its Contra allies, the Agency admitted only one such connection. In 1984, the CIA cut its ties to one Contra leader linked to the notorious cocaine trafficker Jorge Morales but, in the words of the report, "continued to have contact through 1986–87 with four of the individuals involved with Morales." Significantly, the Inspector General's report now reveals that the CIA not only worked with fifty-eight Contras implicated in cocaine trafficking, but also concealed their criminal activities from Congress while the operation was in progress. "In six cases," the report states, "CIA knowledge of allegations or information indicating that organizations or individuals had been involved in drug trafficking did not deter their use by CIA. In at least two of those cases, CIA did not act to verify drug trafficking allegations or information even when it had the opportunity to do so." In 1988, Senator John Kerry and his staff investigators had very limited success in pressing the CIA's chief liaison with Congress, John Helgerson, for information on alleged Contra drug activity. In his investigation, the Inspector General found the reason for this reticence. "Helgerson says that he does not recall a concerted effort by the agency to get to the bottom of the allegations of narcotics trafficking," the report says. "The agency, as he puts it, was in the intelligence business, not law enforcement."[66]

With its lengthy quotations from CIA memorandums and interviews, the Hitz report allows, for the first time, a close analysis of how the Agency's covert warfare fostered enforcement-free zones effectively closed to anti-narcotics investigations. Buried in volume two's back pages, at paragraph number 913, Hitz explores the Agency's alliance with Alan Hyde, a Honduran drug dealer, providing a revealing case study of the operational pressures that led CIA field officers into a compromised relationship with a notorious cocaine trafficker. As so often happens in covert warfare, geography circumscribed the CIA's choices. Lying along the main smuggling routes between Colombia and America's Gulf Coast, the Bay Islands, just off Honduras, were ideally sited for both cocaine smuggling into the United States and CIA arms shipments to the frontline Contra bases along the Honduras-Nicaragua border.

During the 1980s, Alan Hyde, a local businessman, used his fishing fleet and fish-processing plant at Roatan Island to emerge, in the words of the U.S. Coast Guard, as the "godfather for all criminal activities originating in Bay Islands." In 1984, the U.S. Defense Department attaché at Tegucigalpa reported that Hyde "is making much money dealing in 'white gold,' i.e. cocaine." Two years later, the U.S. Coast Guard cited "reliable sources" to report that an "organized criminal organization" led by Hyde was smuggling cocaine from the Bay Islands on "fishing vessels bound for South Florida." At the same time, the Coast Guard advised the CIA that Hyde Shipping of Miami had purchased fishing vessels in Brownsville, Texas, that had sailed to Cartagena, Colombia, "to be outfitted for drug smuggling." Indeed, in March 1988 the CIA's own Directorate of Intelligence issued an assessment titled "Honduras: Emerging Player in the Drug Trade," reporting that Hyde was smuggling chemicals for two cocaine-processing plants in the Bay Islands.

Nonetheless, the CIA relied upon Hyde for logistical support to the Contras from 1987 to 1989, a period when his reputation for drug dealing was already well established. Weighing the covert operational gains against compromising ties to a major drug dealer, Agency personnel at all levels, in an extreme instance of mission myopia, crafted an exculpatory logic that allowed them to deal with Hyde. In July 1987, a CIA field operative first suggested that "using Hyde's vessels to ferry supplies would be more economical, secure and time efficient than using aircraft." At CIA headquarters in Langley, the Central American Task Force (CATF) warned that Hyde "might have ties to drug traffickers," but still permitted ongoing contacts pending higher approval of this larger logistical alliance. A month later, Alan Fiers, head of the CATF, advised his superior, Deputy Director for Operations Clair George, that Hyde was "not attractive" but that the Agency "had no choice but to use Hyde on the grounds of 'operational necessity.'" George, in turn, discussed Hyde with the CIA's deputy director, Robert Gates, saying, "We need to use him, but we also need to figure out how to get rid of him." After Gates approved the relationship at the highest level, George, in a cable dated August 8, 1987, issued the authorization to use Hyde to "provide logistical services to complete a project, after which all contacts must cease."

Over the next two years, CIA headquarters sent what one agent called "mixed signals" about Hyde to its field operatives, expressing growing reservations about his reputation but simultaneously exerting pressure to get the job done. Left to resolve these contradictory directives on his own, one CIA logistics officer "never believed the drug allegations against Hyde, whom he came to regard as a close friend." Since Hyde had thirty-five ships at sea, this officer did admit that "it might have been possible for an employee of Hyde to use one of the boats for smuggling." Another CIA operative who worked with Hyde dismissed all the U.S. government reports about his trafficking, saying, "If these allegations were true, DEA and the Coast Guard would have got him."

Though initial reports of his drug dealing sparked no reaction from CIA headquarters, allegations that Hyde was stealing arms destined for the Contras prompted the Agency's Nicaragua Operations Group (NOG) to dispatch investigators. "It was an issue of relative order of priorities," the former head of the NOG recalled, in words that seemed to epitomize the Contra operation's pervasive mission myopia. "[T]he issue of stealing our guns had a higher priority than verifying whether Hyde was smuggling cocaine into the U.S." Indeed, in March 1988, CIA headquarters authorized the leasing of additional storage facilities from Hyde despite recent internal information that he was "the head of an air smuggling ring with contacts in the Tampa/St. Petersburg area." Within several months, however, "the receipt of continuing allegations of Hyde's involvement in cocaine trafficking" placed pressure on CIA headquarters to find an alternative, particularly once a resumption in Contra aid operations, after a brief suspension, made the Agency less dependent upon his storage facilities to "mothball" supplies. Throughout, field operatives resisted pressure from headquarters to sever ties, arguing that Hyde was their only reliable "delivery mechanism." By early 1989, however, these logistics problems had apparently been resolved, and headquarters finally ordered the field to break off contact.

The CIA alliance apparently shielded Hyde's drug smuggling from serious investigation by either Honduran or American authorities. When the Senate Standing Committee on Intelligence had questioned the CIA's alliance with Hyde in August 1987, Fiers, head of the CATF, had replied that "persons such as Hyde could be used in the Contra program if there were no ongoing investigations of wrongdoing or no outstanding indictments." If we are to take Fiers at his word, both the DEA and Coast Guard, despite past intelligence about Hyde's drug smuggling into the United States, had suspended their investigations now that he was working with the CIA. Six months later, a CIA officer advised headquarters that Honduran authorities planned a "drug-busting effort" against the Bay Islands that would target Hyde, since he was "the 'godfather' of all criminal activities—[especially] drug smuggling—in the Islands." But the Agency's records show no record of any Honduran government raids, raising the possibility that the CIA may have directly or indirectly discouraged the planned operation. Even five years after the Agency finally cut relations, the CIA's apparent efforts to protect Hyde and close his island haven to external investigation continued. In March 1993, an internal directive stated frankly that it was Agency policy to "discourage ... counternarcotics efforts against Alan Hyde because 'his connection to [the CIA] is well documented and could prove difficult in the prosecution stage.'"[67]

In effect, the CIA, understanding the potential sensitivity of drug trafficking by its Central American assets, had apparently formalized its policy of closing covert war zones to external investigation. Remoteness had

Mission Myopia 139

allowed the CIA simply to ignore drug trafficking by its allies in Laos and Pakistan, but Contra involvement in cocaine smuggling into the United States required a more formal waiver to protect the Agency from later charges of illegality. Through his inquiry, Inspector General Hitz established a well-documented case study that corroborated the Agency's long-standing policy of exempting its covert assets from any investigation of drug trafficking done in connection with CIA operations.

Conclusion

During the Cold War, CIA agents allied with ethnic warlords in the Asian opium zone to wage covert wars that served as catalysts for the region's opium traffic at several critical points in its postwar history. In mountain ranges along the southern rim of Asia—whether in Afghanistan, Burma, or Laos—opium was the main currency of external trade and thus a key element in the apparatus of local power. Once allied with the CIA, these highland leaders used the Agency's arms, logistics, and protection to expand both local opium production and their share of its profits, quickly evolving from minor ethnic warlords into powerful drug lords. As Agency assets, leaders such as Hekmatyar and Vang Pao were immune to investigation and could take control of the local opium trade with impunity.

Focused on their covert mission, CIA operatives usually ignored drug dealing by their assets. Once the CIA had identified itself with one of these opium warlords, it could not afford to compromise an important covert-action asset with drug charges. Respecting the national-security imperatives of CIA operations, the DEA kept its distance from Agency assets—whether in Laos, Afghanistan, or Honduras. Such implicit tolerance allowed covert war zones to become enforcement-free areas where opium trafficking could expand without restraint.

Critics who look for evidence that CIA agents have actually dirtied their hands with drugs in the line of duty are missing the point. Since the Agency's operations were covert, it avoided any direct involvement in combat, working instead through local clients, whose success usually determined the outcome of the Agency's operation. The Agency's involvement in drugs took the form of tolerance of drug dealing by covert assets, not, in most instances, direct culpability.

From a narrow Cold War perspective, such informal sanctioning of drug dealing often amplified the CIA's operational effectiveness. In Burma, Laos, and Afghanistan, increased opium production provided critical support for covert operations. As tribal societies mobilized to fight the CIA's secret wars, they diverted essential manpower from subsistence agriculture to combat. Over time, casualties further reduced the food harvest, making the income from opium an important substitute for lost labor. In effect, the diversion of scarce tribal manpower from subsistence farming was covered by a rapid increase in cash-crop drug sales. From the CIA's viewpoint, narcotics income spared the Agency the high—perhaps prohibitively high—cost of providing a full welfare system for a tribe with dependents numbering in the hundreds of thousands. Of equal importance, control over this critical source of cash income allowed the CIA's leaders to command tribes, clans, and villages in bloody paramilitary campaigns that ground on for a decade or more. In sum, through its tolerance of the opium trade, the CIA's expenses declined and its operational effectiveness increased.

Looking back on the major CIA paramilitary campaigns that became enmeshed in narcotics traffic, there is a striking contrast between the short-term gains in paramilitary capacity and the troubling, long-term legacy of these political compromises. During each operation and its aftermath, there was a sharp rise in both opium production and opiate exports to international markets, including Europe and America. During the 1950s, when the CIA allied with Nationalist Chinese irregulars in northern Burma, the rising opium harvest contributed to the region's later emergence as opium's Golden Triangle. During the 1960s, the CIA's Laotian assets opened the region's earliest heroin refineries, exporting first to American troops fighting in South Vietnam and then following the GIs home to capture a quarter of the U.S. drug market. In the Afghan war of the 1980s, the CIA's regional assets exploited its covert operation to make the Pakistan-Afghanistan border into the world's largest source of illicit heroin. Simultaneously, CIA operatives in Central America, many of them veterans of these Asian campaigns, tolerated the involvement of their regional allies in the cocaine traffic that passed northward to the United States. In this case, the CIA's decision to sacrifice the drug war to fight the Cold War created an image of insensitivity to the rising inner-city crack plague that would later erupt in racially charged controversy.

Once the CIA's covert wars ended, their legacy persisted in steadily rising narcotics production. In Burma, Laos, and Afghanistan, opium and heroin production continued to increase in the decades after the Agency's departure. By the time the CIA withdrew the logistic and political support that had been instrumental in the traffic's early expansion, the local drug trade had already achieved an economic "takeoff point" that allowed sustained, independent growth. The American agents may have departed, but the logistics, market linkages, and warlord power remained to make these regions major drug suppliers for decades to come. Of equal importance, the peculiar character of covert warfare denied its battlegrounds the postwar reconstruction aid that often follows conventional combat. Since these

Mission Myopia 141

secret wars are fought outside normal diplomatic channels, their postwar rehabilitation often remains beyond the power of bilateral or multilateral institutions. In lieu of international aid, these highland societies have expanded their opiate production as an ad hoc form of postwar reconstruction. Over time, narcotics production not only serves to sustain a traumatized society during its postwar recovery, but also reinforces its isolation from the legitimate economic and political resources of the international community. Though small and remote, these highland societies have, in the aftermath of these covert wars, become significant sources of international instability—wounds on the international body politic.

With this background, we can now return to the four questions posed at the outset.

Was the CIA ever allied with drug traffickers? Yes, beyond any doubt. Though once a controversial charge, not even the Agency any longer bothers to deny that it has often formed tactical alliances with drug dealers in Asia and the Americas.

Did the CIA protect these allies from prosecution? Yes. There is a recurring pattern of de facto protection. During a major CIA operation, the covert war zone becomes a special "protected area" in which everything is subordinated to the prosecution of the mission. To insure secrecy, the CIA blocks all investigations—by the DEA, Customs, Congress, and the police. In the case of the Contra base camps, where the CIA was the only U.S. investigative agency, it won a waiver of any responsibility for reporting on the criminal activities of its assets. In effect, the Agency's key allies have been given a de facto immunity to prosecution.

Did the CIA encourage drug smugglers to target African-American communities? There is no evidence or logic that would suggest any such targeting. The pattern of CIA complicity in drugs, described above, proceeds from the internal logic of its covert operations, a consequence of its tactics of indirect intervention through paramilitary operations. There is a striking similarity in the patterns of CIA complicity with drug dealers in Laos, Afghanistan, and Central America. Just as there is little evidence and less logic to the proposition that the CIA wanted one-third of the U.S. troops fighting in South Vietnam to become heroin addicts, there is no evidence that the CIA targeted blacks in Los Angeles. During the 1980s, however, there was every indication that the CIA was aware that its Afghan and Central American operations contributed to the export of cocaine and heroin to the United States—and did nothing to slow this drug flow.

Finally, have the CIA's alliances with drug lords contributed significantly to an expansion of the global drug trade over the past forty years? In Burma, Laos, and Afghanistan, CIA operations provided critical elements—logistics, arms, and political protection—that were catalytic in the rapid growth of opium and heroin production. It would not be unreasonable to conclude, therefore, that such CIA operations led to an increase in the production and processing of illicit drugs in these covert war zones. But it is difficult to state unequivocally that the individual drug lords allied with the CIA did or did not shape the long-term trajectory of supply and demand within the vastness and complexity of global drug traffic.

Whatever the global effect of CIA covert warfare, the Agency's alliances with drug lords have left, in the aftermath of the Cold War, a domestic legacy of illegality, suspicion, and racial division. Suffering from mission myopia, CIA agents fighting secret wars in Laos, Pakistan, and Central America seemed to regard narcotics as mere "fallout," even when the victims were U.S. soldiers in South Vietnam or Americans in the inner cities. By ignoring the drug trafficking of its covert allies, the CIA violated federal laws and later executive orders requiring it to provide intelligence in support of the drug war. In a society that seeks to perfect the rule of law, such willful illegality has lasting repercussions. Indeed, among many African-Americans the Agency's apparent callousness toward the victims of these covert compromises has sparked deep suspicions of institutional racism that will take decades to repair.

The history of CIA complicity also raises some important questions about the Agency's future role in U.S. foreign policy. At the broadest level, this catalog of the CIA's criminal alliances reveals a great deal about the moral and political costs of investing an executive agency with extraordinary powers—questions that this society has refused to address in a serious, sustained manner.

As winner of the Cold War, the United States has been spared any painful self-examination, any need to question the methods we used or the price we paid for victory in the longest of our nation's wars. In contrast, our former enemies, the people of Russia and Eastern Europe, have been forced into agonizing self-reflection as they try to rebuild societies ravaged by decades of authoritarian rule. Through these struggles for reform and renewal, communism's notorious espionage agencies—the Stasi, Securitate, and KGB—have been swept away, their files opened, and in some cases their leaders investigated and even indicted. Alone among the major covert agencies that fought the Cold War, the CIA survives, its files still sealed, its crimes still largely unexamined. Not only has the Agency survived, but it has used its claims of victory over and ability to contain communism to win recent budget increases that make it one of the most powerful of federal agencies.

We need to put aside the comfortable self-assurance stemming from our Cold War victory and begin asking some hard questions about the CIA's future. Now that the Cold War is over, do we really need a CIA armed with the extraordinary powers that place it beyond the law? If so, do we want the CIA to fight the drug war with the same covert-action arsenal it used in the Cold War?

On its fiftieth anniversary in November 1997, the CIA looked at a future without communism and proclaimed its new mission to be the fight against international crime, particularly drug trafficking. As President Clinton put it, the Agency would be charged with "protecting American citizens from new transnational threats such as drug traffickers, terrorists, organized criminals, and weapons of mass destruction."[68] Yet there is good reason to doubt the CIA's usefulness in the war on drugs. As an intelligence, espionage, and covert-action agency, the CIA has developed an ingrained institutional culture of operating outside the law. Unless we are to adopt President Nixon's option of wholesale assassination of drug lords, the CIA is ill-equipped for the fight against crime. Unlike the FBI or DEA, it simply does not have the experience to collect evidence within the law in ways that will allow successful criminal prosecutions. Moreover, as shown in its relations with the Nationalist Chinese and the Contras, the CIA is compromised by its past alliances with drug lords. Whenever it has been confronted with an exposé of drug trafficking by its former allies, the Agency has gone to extraordinary lengths to cover the evidence. In 1972, for example, the CIA paid the Nationalist Chinese irregulars in northern Thailand nearly $2 million to burn their "last" twenty-six tons of opium—a sham publicity spectacle staged to erase the Agency's long alliance with Asia's leading drug lords.[69] Clearly, the CIA is extraordinarily sensitive about its past complicity and may find itself compromised if it attempts to investigate its underworld allies.

Above all, we need to ask whether we want the CIA to preserve its Cold War powers to conduct covert operations, exempt from both legal restraint and legislative oversight. There is reason to believe that the CIA's reliance on criminals for these special missions was not just a product of the Cold War. Judging from its history, the CIA often needs criminals and their special skills to carry out its covert operations, whether the goal is the destabilization of a hostile regime, the assassination of a foreign leader, or the mobilization of a mercenary army. The alliances with drug dealers and drug lords were not an aberration, an expedient born of the Cold War. To mount complex covert operations in remote foreign territories, Agency operatives often seem to require the services of assets skilled in what one CIA agent, Lucien Conein, called "the clandestine arts": the ability, shared by spies and criminals, to conduct vast enterprises or major operations outside the bounds of civil society in ways that leave no trace. Simply put, criminal liaisons are an integral part of the CIA's covert operational capacity.

When the White House orders future covert operations, as opposed to routine espionage, the Agency will still need criminals who can operate in foreign cities or local warlords who can mobilize mercenaries in the countryside. In either case, such allies will, in all probability, use the CIA's protection to traffic in drugs. As we saw in the Contra operation, past regulations barring the CIA from criminal alliances have not proved effective. In sum, if the executive branch orders the CIA to carry out covert operations, the Agency's criminal associations and tolerance of drug dealing are likely to recur.

Nearly a decade after the end of the Cold War, we seem to be faced with a clear choice. We can either deny the Agency the authority to conduct covert operations, or we can accept that these missions will involve the CIA in criminal alliances that may well compromise some future war on drugs. There is, of course, a moral dimension to this choice. Every nation needs an intelligence service to warn of future dangers. But a nation at peace has no right, under U.S. or international law, to conduct its foreign policy through covert operations involving bribes, black propaganda, murder, or undeclared warfare.

NOTES

1. *San Jose Mercury News,* August 18–20, 1996.
2. *San Jose Mercury News,* August 21, 1996.
3. *Boston Sunday Globe,* October 6, 1996; *New York Times,* October 21, 1996; *Washington Post,* September 13, 1996; Peter Kornbluh, "The Storm Over 'Dark Alliance,'" *Columbia Journalism Review,* January/February 1997, pp. 33–35.
4. *Washington Post,* September 13, 1996; Letter from Representative Maxine Waters to U.S. Attorney General Janet Reno, August 30, 1996.
5. *Washington Post,* October 4, 1996; *New York Times,* October 21, 1996; *Los Angeles Times,* October 20–22, 1996.
6. *New York Times,* November 14, 1993; April 22, 1994; June 8, 1994; March 3, 1997; July 17, 1998; *Dallas Morning News,* December 1, 1996.
7. This history of the colonial opium trade is analyzed at length in Chapter 3 of my book *The Politics of Heroin: CIA Complicity in the Global Drug Trade* (New York: Lawrence Hill Books, 1991).
8. League of Nations, "Advisory Committee on the Traffic in Opium and Other Dangerous Drugs, Annual Reports on the Traffic in Opium and Other Dangerous Drugs for the Year 1939" (Geneva, Switzerland: League of Nations, 1940), p. 42.; Circular No. 875-SAE, July 22, 1942, from Resident Superior of Tonkin Desalle, to the residents of Laokay, Sonla, and Yenbay, quoted in *Association culturelle pour le Salut de Viet-Nam, Temoinages et Documents français relatifs a la Colonisation française au Viet-Nam* (Hanoi, 1945), p. 115.

9. A detailed, documented analysis of the CIA's Burma operation can be found in *The Politics of Heroin,* Chapter 4.

10. Chao Yawnghwe, *The Shan of Burma: Memoirs of a Shan Exile* (Singapore: Institute of Southeast Asian Studies, 1987), p. 57; U.S. Cabinet Committee on International Narcotics Control, *World Opium Survey 1972* (Washington, DC: CCINC, 1972), pp. A35–A37.

11. An analysis of French and American complicity in the Indochina opium trade can be found in *The Politics of Heroin,* Chapters 4 and 7.

12. Senate Select Committee to Study Governmental Operations with Respect to Intelligence Activities, 94th Congress, 2nd Session, *Foreign and Military Intelligence,* Book I: *Final Report* (Report No. 94-755) (Washington, DC: Government Printing Office, 1976), pp. 228, 232–233.

13. *Foreign and Military Intelligence,* Book I, p. 229.

14. U.S. Executive Office of the President, Special Action Office for Drug Abuse Prevention, *The Vietnam Drug User Returns: Final Report* (Washington, DC: Government Printing Office, 1974), p. 57.

15. "Turkey Lifts the Poppy Ban" by John T. Cusack, Director, International Operations Division, DEA Office of Enforcement, in *Drug Enforcement,* Fall 1974, pp. 3–7; House Committee of Foreign Affairs, 95th Congress, 1st Session, *Foreign Assistance Legislation for Fiscal Years 1984–85 (Part 5): Hearings and Markup Before Subcommittee on Asian and Pacific Affairs* (Washington, DC: Government Printing Office, 1983), pp. 556–557.

16. *Imperial Gazetteer of India: Provincial Series, North-West Frontier Province* (Calcutta: Superintendent of Government Printing, 1908), p. 25; Barnett R. Rubin, "The Fragmentation of Afghanistan," *Foreign Affairs* 68, No. 5 (1989): 150–168; C. Colin Davies, *The Problem of the North-West Frontier, 1890–1908* (London: Curzon Press, 1932), pp. 24–26.

17. *Report on the Administration of the Punjab and Its Dependencies for the Year 1870–71* (Lahore: Government Civil Secretariat Press), p. cxxxiii; *Imperial Gazetteer of India: North-West Frontier Province,* pp. 65–66.

18. Cabinet Committee on International Narcotics Control, *World Opium Survey* (1972), p. A-11.

19. Garland H. Williams, "Opium Addiction in Iran," report to H. J. Anslinger, Commissioner of Narcotics (February 1, 1949), pp. 1–12 (Historical Collections and Labor Archives, Pennsylvania State University).

20. CIA Directorate of Intelligence," Intelligence Memorandum: Narcotics in Iran" (part of International Narcotics Series No. 13) (Washington, DC: Central Intelligence Agency, June 12, 1972), pp. 2–3; Harry J. Anslinger, Letter to William J. Stibravy, Director, Office of International Economic and Social Affairs, United Nations (October 28, 1968) (Historical Collections and Labor Archives, Pennsylvania State University).

21. CIA, "Intelligence Memorandum: Narcotics in Iran," pp. 2–10.

22. Richard Helms, Airgram From: Amembassy Tehran, To: Department of State, Subject: Revised Narcotics Action Plan, Date: March 4, 1974 (Department of State).

23. Richard Helms, From: Amembassy Tehran, To: SecState Wash DC 2222, Subj: Narcotics Matters Related to Iran, R: 261315Z. June 1975 (Department of State).

24. Thomas T. Hammond, *Red Flag over Afghanistan* (Boulder, CO: Westview Press, 1984), pp. 98–99, 118–121, 158–159.

25. Tariq Ali, *Can Pakistan Survive?* (New York: Penguin, 1983), pp. 133–140; Hammond, *Red Flag over Afghanistan*, pp. 218–221; Lawrence Lifschultz, "Dangerous Liaison: The CIA–ISI Connection," *Newsline* (Karachi), November 1989, p. 52.

26. House Foreign Affairs Committee, Barnett R. Rubin: Testimony before the Subcommittee on Europe and the Middle East and the Subcommittee on Asia and the Pacific. "Answers to Questions for Private Witnesses" (March 7, 1990), pp. 3–4; Barnett R. Rubin, "The Fragmentation of Afghanistan," *Foreign Affairs* 68, No. 5 (1989): 154.

27. Barnett R. Rubin, Testimony before the Subcommittee on Europe and the Middle East (March 7, 1990), p. 5.

28. Lifschultz, "Dangerous Liaison," pp. 49–54.

29. U.S. Department of State,. "Pakistan's Tribal Areas: How They Are Administered or Not" (October 1986). Cited in: Lucy Mathiak, "Unholy Alliance: The War on Drugs and Our Drug Lord Allies" (draft essay, 1998), p. 10.

30. CIA, Directorate of Intelligence, "Passes and Trails on the Pakistan-Afghanistan Border: A Reference Aid." (April 1983). Cited by Lucy Mathiak in "Unholy Alliance," p. 10.

31. Lifschultz, "Dangerous Liaison," pp. 52–53.

32. Lawrence Lifschultz, "Inside the Kingdom of Heroin," *The Nation*, November 14, 1988, pp. 492–493.

33. Lifschultz, "Dangerous Liaison," pp. 52–53.

34. Malthea Falco, Director for International Narcotics Control Matters, U.S. Department of State, "Asian Narcotics: The Impact on Europe," *Drug Enforcement*, February 1979, pp. 2–3; Cabinet Committee on International Narcotics Control, *World Opium Survey 1972*, pp. A-7, A-14, A-17; Department of State, *International Narcotics Control Strategy Report* (Washington, DC: U.S. Government Printing Office, February 1, 1984), p. 4; William French Smith, "Drug Traffic Today—Challenge and Response," *Drug Enforcement*, Summer 1982, pp. 2–3.

35. Department of State, *International Narcotics Control Strategy Report*, p. 4.

36. *New York Times*, June 18, 1986.

37. Kathy Evans, "The Tribal Trail," *Newsline* (Karachi), December 1989, p. 30.

38. Rubin, Testimony before the Subcommittee on Europe and the Middle East, pp. 18–19; *Washington Post*, May 13, 1990.

39. Evans, "The Tribal Trail," p. 26.

40. Lifschultz, "Inside the Kingdom of Heroin," pp. 495–496.

41. Pakistan Narcotics Control Board, *National Survey on Drug Abuse in Pakistan* (Islamabad: Narcotics Control Board, 1986), pp. iii, ix, 23, 308; Zahid

Hussain, "Narcopower: Pakistan's Parallel Government?" *Newsline* (Karachi), December 1989, p. 17.

42. House Committee on Foreign Affairs, 95th Congress, 1st Session, *Foreign Assistance Legislation for Fiscal Years 1984–85,* pp. 229, 234–235, 312–313, 324–325, 326–327, 500–504, 513, 528–529, 544–545, 548–550.

43. Hussain, "Narcopower," p. 17; Lifschultz, "Inside the Kingdom of Heroin," pp. 492–495; Lawrence Lifschultz, "Turning a Blind Eye?" *Newsline* (Karachi), December 1989, p. 32.

44. General Accounting Office, Drug Control: U.S. Supported Efforts in Burma, Pakistan, and Thailand (Washington, DC: Report to Congress, GAO/NSIAD 88-94, February 1988), pp. 25–34, cited by Lucy Mathiak in "Unholy Alliance," p. 10; Lifschultz, "Inside the Kingdom of Heroin," pp. 495–496.

45. *New York Times,* September 7, 1988.

46. Hussain, "Narcopower," pp. 14-a, 17.

47. Evans, "The Tribal Trail," p. 24.

48. Department of State, Bureau of International Narcotics Matters, *International Narcotics Control Strategy Report* (Washington, DC: U.S. Department of State, March 1, 1990), p. 239.

49. Rubin, Testimony before the Subcommittee on Europe and the Middle East, pp. 18–19.

50. Rubin, Testimony before the Subcommittee on Europe and the Middle East, pp. 20, 35.

51. Barnett R. Rubin, The Situation in Afghanistan: Testimony before the Commission on Security and Cooperation in Europe, U.S. Congress (May 3, 1990), pp. 16–18.

52. *Washington Post,* May 13, 1990.

53. *Time* (Australian edition), July 16, 1990, pp. 28–29.

54. Aspire Films, *Dealing With the Demon: Part II* (produced by Chris Hilton) (Sydney, 1994).

55. Brian Barger and Robert Perry, Associated Press, December 20, 1985.

56. Senate Committee on Foreign Relations, Subcommittee on Terrorism, Narcotics and International Operations, 100th Congress, 2nd Session, *Drugs, Law Enforcement and Foreign Policy* (Washington, DC: Government Printing Office, December 1998), pp. 26–29, 39–54.

57. *Drugs, Law Enforcement and Foreign Policy,* p. 75.

58. *Washington Post,* December 7, 1987.

59. *San Jose Mercury News,* August 18–19, 1996.

60. *Los Angeles Times,* October 8, 1996 and December 11, 1996; *San Francisco Chronicle,* October 8, 1996; *Sacramento Bee,* October 8, 1996.

61. *Los Angeles Times,* December 30, 1996; Statement of Frederick P. Hitz, Inspector General Central Intelligence Agency Before the Permanent Select Committee on Intelligence United States House of Representatives, March 16, 1998, pp. 1–2.

62. *San Jose Mercury News,* December 19, 1997; *New York Times,* January 30, 1998.

63. *Washington Post,* March 17, 1998; Alexander Cockburn and Jeffrey St. Clair, *White Out: The CIA, Drugs and the Press* (New York: Verso, 1998), pp. 385–392; Martha Honey, "Don't Ask, Don't Tell," *In These Times,* May 17, 1998.

64. *New York Times,* July 17, 1998.

65. *New York Times,* October 10, 1998.

66. John Diamond, "CIA Admits Not Informing on Contras" (Washington, DC: AP Online, Lexis-Nexis, October 10, 1998); Alexander Cockburn, "CIA's Trail Leads to its Own Door," *Los Angeles Times,* October 22, 1998.

67. U.S. Intelligence Agency, Office of the Inspector General, *Allegations of Connections Between CIA and the Contras in Cocaine Trafficking in the United States* (1) (96-0143-IG). Vol. II: *The Contra Story,* paragraphs 913–961.

68. Cockburn and St. Clair, *White Out,* p. 393.

69. House Select Committee on Narcotics Abuse and Control, 95th Congress, 1st Session, *Southeast Asian Narcotics* (Washington, DC: Government Printing Office, 1978), pp. 178–179; Jack Anderson, "Thai Opium Bonfire Mostly Fodder," *Washington Post,* July 31, 1972, p. B-11.

Robert Dreyfuss

8 TECHINT
The NSA, the NRO, and NIMA

DWARFING THE CENTRAL INTELLIGENCE AGENCY in the U.S. intelligence community is a constellation of agencies devoted to technical intelligence, or TECHINT. Centered in the National Security Agency (NSA), the National Reconnaissance Office (NRO), and the National Imagery and Mapping Agency (NIMA), and working closely with the military intelligence services, these agencies operate a vast network of surveillance satellites, spy planes, and electronic listening posts on land and under the sea worth well over $100 billion.[1] Together, the three agencies employ more than 30,000 people, sustain an impressive array of corporate partners who serve as contractors, and share an annual budget estimated at $11 billion a year,[2] roughly 40 percent of what America spends on intelligence and nearly four times the budget for the CIA.

Despite their enormous size, however, the U.S. technical-intelligence agencies are not well known to the public. Their obscurity is intentional, since the nature of their work is highly classified and closely guarded by a priesthood of scientists, engineers, computer technicians, and mathematicians. The agencies have been able to operate behind a cloak of secrecy for another reason: Unlike the CIA, which attracts attention for its network of spies and its exploits in paramilitary and covert actions, the technical-collection agencies operate silently in space and in the bowels of the world's telecommunications systems. Yet the intelligence provided over the years by the NSA and the NRO has formed the very backbone of the information available to the U.S. intelligence community and to the analysts at the CIA and the Pentagon. The TECHINT agencies have provided American officials with a steady and reliable stream of intelligence, including images of Soviet nuclear-missile launch sites during the Cold War and transcripts of private communications of countless foreign leaders, both friend and foe.

Like the CIA, the NSA and NRO are creatures of the Cold War. The NSA was established in 1952 and the NRO in 1960, at the height of the worldwide struggle between the United States and the Soviet Union. For most of their existence, their satellites and communications gear were directed against the USSR and its allies. Though their list of intelligence priorities was long, paramount was the hourly tracking of the Soviet Union's armed forces and its nuclear arsenal, especially during periods of crisis in such areas as Cuba and the Middle East, when America's very existence was conceivably at risk from thermonuclear war. In the early 1960s, the fledgling NRO's photographs of Soviet missile sites helped to dispel allegations of a "missile gap." Then, as détente warmed, the NRO played a critical role by giving U.S. officials the ability to verify Soviet compliance with arms treaties. In the 1990s, however, there are questions about whether the NSA and NRO have adapted to the end of the Cold War. Unlike the Pentagon, whose size has been significantly reduced since the mid-1980s, the technical-intelligence agencies have apparently entered the post–Cold War era with their budgets largely intact.

Whether these agencies can be reorganized and downsized for the twenty-first century is a question worth asking. However, evaluating the NSA and the NRO is extremely difficult for outsiders (that is, the public) because just about everything they do—and, until recently, the NRO's very existence—is classified. Despite the huge sums involved, and despite the awesome power placed in the hands of these agencies to pry into any corner of the world and into virtually all forms of electronic communications, few details about their operations have been released to the public. Still, a vigorous public debate is required about the mission of TECHINT in a world without the Soviet adversary. Are the same resources and the same level of expenditure required to monitor the potential proliferation of nuclear weapons and other weapons of mass destruction and the activities of secondary nuclear powers as were required to keep track of the USSR's threat to the very existence of the United States? And what about TECHINT's role in other missions, such as economic intelligence, environmental trends, counterterrorism, drug trafficking, and support to military operations?

Though the budget numbers are secret, answers to these questions are even more important because during the 1990s, both the NRO and the NSA have been accused of poor financial management and wasteful spending.[3] Evaluation by the U.S. government, necessarily behind closed doors, occurs through two mechanisms. First, theoretically at least, the CIA director, in his role as chief executive of the U.S. intelligence community, is responsible for the management of all three agencies—but in fact, they are quasi-independent fiefdoms with their own directors and separate budgets included within the "black" budget of the Department of Defense. Second, Congress (specifically the House and Senate Intelligence Committees) performs a limited oversight function—but the complexity of the agencies' missions, the layers of secrecy that surround them, and the lobbying power of the intelligence-industrial complex raise doubts about how well the legislative branch is able to oversee them.

The TECHINT missions of America's spy agencies fall roughly into two broad categories: imagery (IMINT) and signals intelligence (SIGINT). "Imagery" is a catchall term that includes everything from old-fashioned aerial photography to sophisticated radar images. Signals intelligence includes two subcategories: The first is communications intelligence, or COMINT, which involves the interception and processing of a wide range of electronic communications, including telephone calls, modem and fax transmissions, and radio communications involving governments, businesses, and individuals. The second is ELINT, or electronic intelligence, useful mostly for military purposes. ELINT is defined as "the collection (observation and recording) and the processing for subsequent intelligence purposes, of information derived from foreign non-communications electromagnetic radiations."[4]

The National Security Agency is chiefly responsible for SIGINT.[5] Based at Fort Meade, Maryland, between Washington, D.C., and Baltimore, the NSA sprawls over a 650-acre campus that includes 7 million square feet of office space, its own 400-member police force, a power plant, a printing facility, and a computer-chip factory to supply the agency's computers.[6] The NSA's 21,000 employees make it the largest employer in Maryland, with an annual payroll of $930 million.

A key NSA unit is the Central Security Service, responsible for codebreaking—that is, deciphering other nations' encrypted communications—and for overseeing U.S. communications and information security systems, including secure links for the CIA, the Federal Bureau of Investigation, the State Department, and the Pentagon. From its headquarters, the NSA runs a worldwide empire that includes satellite listening devices (run by the NRO), sophisticated eavesdropping aircraft and ships, and key ground stations in West Virginia, Washington State, and Alaska. The agency also operates major foreign listening posts in Australia, Canada, Japan, Germany, and elsewhere, with a total of fifty stations in twenty countries, along with several dozen installations around the world at U.S. embassies, whose rooftops often sport visible arrays of antennae.[7] "They have a dish for every communications satellite," says John Pike, space analyst with the Federation of American Scientists. "Every time a satellite goes up, NSA puts a dish up."[8]

The NSA's chief responsibility is the collection and processing of intelligence derived from intercepted foreign communications. Vast amounts of data are gathered by the NSA's varied collection systems, which have been described as a sort of giant vacuum cleaner for information. Yet the apparatus can be precisely targeted. The NSA can concentrate with great intensity on a particular country or capital, or it can sort through its worldwide communications intercepts for mentions of particular names, words, or phrases. Should one of the NSA's customers at the CIA, the Pentagon, or elsewhere require it, particular names or words can be entered into the NSA's "watch list" or "dictionary," and computers will sort through seemingly endless haystacks of information for a few precious needles of data. They will then routinely keep track of all such future references.

The NRO, once the most secret U.S. spy agency, came out of the closet only in 1992. Before that, its existence—even its name—was classified. Representative George Brown of California was kicked off the House Permanent Select Committee on Intelligence in the 1980s just for mentioning the NRO in public remarks. The NRO was established in 1960 to manage the development of America's satellite reconnaissance effort and to coordinate U-2 spy planes and other surveillance aircraft. Since then, with the expansion of space and communications technology, the organization has grown rapidly. Today, the NRO spends roughly $6 billion a year to design, develop, purchase, launch, and manage a vast array of satellites, including the Advanced KH-11 and Lacrosse behemoths, which cost an estimated $1 billion each.

The NRO does not perform analysis itself; it feeds data to the NSA, the CIA (especially the Directorate of Science and Technology), NIMA, and other agencies, where computers and expertly trained human analysts turn raw imagery and SIGINT into finished intelligence. Unlike the labor-intensive NSA, the NRO is hardware-driven and so has a relatively small staff, just 2,190 by one estimate, along with 1,000 on-site contractor personnel.[9] More than any other spy agency, the NRO works intimately with a network of primary and secondary private-sector contractors, led by Lockheed Martin and including TRW, Boeing, Rockwell, E-Systems, and others. According to a former senior CIA official, in 1996 Lockheed Martin alone absorbed more than half of the NRO's entire budget.[10] So lavish is the agency that in 1995 it was pilloried for spending more than $300 million on its new headquarters in the Virginia suburbs of Washington, D.C., without telling Congress, the Department of Defense, or the Director of Central Intelligence. Subsequent audits revealed that the NRO had amassed $4 billion in surplus funds held in a secret account.[11]

The workhorses of the NRO's imaging stable are the Advanced KH-11/Improved Crystal optical systems, code-named Kennan, and the Lacrosse radar-imagery satellites. The Kennan satellites contain a high-resolution mirror telescope, not unlike the one aboard the Hubble space telescope, that can be focused widely or narrowed to laser-like precision. The telescope feeds images to the gadgetry aboard the satellite, which can record both visible light—i.e., photographs, called "electro-optical images"—and invisible (or "infrared") radiation in the form of heat given off by an object. The latter capability allows the satellite to "see" images at night. In addition, the KH-11 carries a device called a photomultiplier tube, which allows the camera to take advantage of ultralow light conditions.

The Lacrosse uses radar to record images. Bouncing signals off the earth and recording the echo, the "synthetic aperture radar" devices can penetrate clouds and atmospheric turbulence to create images in breathtaking detail.

According to the Federation of American Scientists' John Pike, there are at least two KH-11s and two Lacrosses currently in orbit. Sixteen times each day, the satellites orbit the globe, cutting a 1,000-mile-wide swath on each pass and covering roughly the entire planet. The precise capabilities of these satellites are highly classified, though it is widely assumed that the NRO's imagery can be measured in inches of resolution, far better than the one-meter-resolution images that have recently become commercially available. To some degree, the resolution provided by NRO satellites depends on how far away the target is—that is, on how oblique the angle at which the satellite "looks" at the target—but intelligence officials refer half-seriously to the NRO's ability to read license plates from space.

The sleepless eyes in the sky encase their data in code and zap it deep into space, where it is collected by another system of communications satellites. These, in turn, blurt out huge volumes of encrypted data in digital form in seconds. Most of this information is beamed to a bunker-like concrete building in Fort Belvoir, Virginia, which serves as the downlink center. Once on the ground, the satellite data pass through what the NRO regards as its most precious possession, the software that coaxes unworldly clarity and three-dimensional, "virtual reality"-type images from the raw input. By using computers to combine overhead photography and radar imaging from various angles, the wizards at the NRO can generate computer-animated images that make the dinosaurs in *Jurassic Park* look like Gumby.

The third agency in America's TECHINT constellation is the National Imagery and Mapping Agency (NIMA). With headquarters in Bethesda, Maryland, and major facilities in St. Louis, in northern Virginia, and scattered around Washington, D.C., NIMA has an annual budget of $1.2 billion and about 9,000 employees.[12] It was established in 1996 by combining the Defense Mapping Agency, the Central Imagery Office, the Defense Dissemination Program Office, the CIA's National Photographic Interpretation Center, parts of the Defense Airborne Reconnaissance Office, and the imagery-related units of the NRO, the CIA, and the Defense Intelligence Agency. Although it has other missions, NIMA is first and foremost a combat-support agency, part of an overall U.S. effort to achieve "dominant battlespace awareness." Using imagery and information from the entire U.S. intelligence community, NIMA is designed to provide state-of-the-art data to every level of the national-security apparatus, right down to the battlefield commander.

Technology, particularly the digital revolution, is the lifeblood of the NSA, the NRO, and NIMA, but for all three agencies that technology is something of a double-edged sword. From World War II until the 1980s, the NSA and the NRO (or their predecessors) maintained an overwhelming technological advantage over rival spy agencies and the private sector. The NSA operated the world's most sophisticated computers and had a near-monopoly on cryptography. The NRO, from the 1960s onward, developed a series of imagery systems aboard satellites that was unparalleled; only the Soviet Union had anything remotely comparable, and no other country maintained even the pretense of trying to develop competing overhead-reconnaissance systems. Recent advances in space science and computer technology, however, pose serious challenges to the TECHINT agencies. Even as it enhances their capabilities, technology is making their missions more difficult, creating new competitors, placing expensive new demands on their product, and raising questions about redundancy and overcapacity.

THE TECHNOLOGICAL CHALLENGE TO THE NSA AND NRO

For the NSA, the proliferation of sophisticated private encryption systems and the use of even more powerful encryption by foreign governments are making it difficult or impossible for the agency's code-breakers to decipher communications by governments and individuals abroad.[13] At the same time, the NSA is struggling to deal with the growing use of optical fiber, which is vastly less vulnerable to interception than earlier avenues of transmission; the ever-increasing speed of data transmission (by faster modems, for instance); and new, complicated switching systems and forms of communications routing. Domestically, both the NSA and the FBI are so worried about the intelligence and law-enforcement problems posed by the spread of high-tech communications that they have persuaded Congress to require telephone companies to modify existing and future telecommunications networks in order to accommodate NSA and FBI wiretaps or other intercepts, and to appropriate $500 million to cover the companies' costs of doing so.

Yet overseas, the NSA has no control over what public or private telecommunications systems do, and it is not certain that even with its supercomputers and tens of thousands of scientists, engineers, and cryptographers, the NSA will be able to unravel the astronomically more complex encryption now available. In the private sector, less powerful encryption systems can be cracked fairly easily by computer scientists. For instance, in July 1998, a small team of experts using a homemade computer costing less than $250,000 broke the so-called Data Encryption Standard, which is widely used by banks and other financial institutions.[14] The financial world would no doubt welcome a code that no one could break, but for an agency whose raison d'être is cracking codes, the generalized use of unbreakable ones is a nightmare scenario.

How has the spread of communications technology affected the NSA? No one knows outside Fort Meade, and it is by no means clear that the NSA is completely prepared to cope with the ongoing technological revolution. In 1996, the Brown Commission admitted that "evaluating the 'technological health' of NSA exceeded the capability of this Commission," and recommended that "an objective and systematic evaluation of this subject be undertaken in order to decide what personnel skills are needed, where research and development should be concentrated, and where investments should be made."[15] One troubling sign: The commission reported that, reflecting the ever-greater costs of hiring skilled scientists and technicians, the NSA's civilian payroll had increased from 30 percent of its total budget in 1990 to 40 percent in 1996.[16]

The NSA's problems were also noted by the report of the House Committee on Intelligence: "Signals intelligence today is at a crossroads. The global revolution in communications technology demands new techniques, new procedures, and a new corporate mindset. The technological challenges currently facing the SIGINT community are daunting...." The report stated that more money would be needed to allow the NSA to keep pace with technology, but noted that "SIGINT is already the most expensive of the intelligence disciplines."[17]

The NRO, too, is severely challenged by the spread of previously closely held spy technology, particularly the emergence of commercial imagery systems. Though virtually all military and intelligence analysts believe that the NRO's imagery product will always be the bedrock of the U.S. intelligence community, the NRO that exists in ten years will probably not resemble the one that exists today. Said the House study, "IMINT will see a great transformation in the next century. Commercial systems will allow everyone, including our foes, to have access to high resolution imagery.... The number of users and requirements will grow. Exploitation will be the chokepoint in the imagery process. The explosion of available imagery will overwhelm the imagery analyst unless automated/assisted target recognition algorithms or other exploitation/production tools can be developed."[18] One effort to deal with the problem of managing IMINT was the creation of NIMA, the result of a consensus among intelligence professionals that imagery analysis, processing, and dissemination should be centralized.

Among the problems facing the NRO is the increased demand for imagery in areas that go far beyond traditional national-security issues. As we shall see, both the NRO and the NSA have seen their list of "customers" in the federal government expand from the CIA, Pentagon, and National Security Council to the Departments of Commerce, Agriculture, and the Treasury; the Environmental Protection Agency; the Federal Reserve; and more. Additional demands are placed on NRO satellites by the interagency committee called COMIREX (Committee on Imagery Requirements and Exploitation). Even as the NRO is pulled by demands for more imagery relating to economics and environmental trends, the Pentagon is demanding that the agency put intelligence directly into the hands of pilots and field commanders on the front lines—a request far beyond the NRO's traditional strategic mission.

But underlying questions about the NRO's effort to redirect its collection toward new targets and new customers is a more fundamental issue: the role of the NRO in the post–Cold War world. The imagery agency is caught in a paradoxical bind. At the precise moment that technology is advancing at a blindingly rapid pace, the targets that preoccupied the NRO for decades—namely the worldwide military forces of the Soviet Union—have largely disappeared. As John Pike of the Federation of American Scientists puts it: "The capability of NRO has increased by an order of magnitude over the past decade, while the number of targets it has to worry about has been reduced by an order of magnitude."[19] The NRO vigorously, but perhaps implausibly, disagrees. "There are more targets today, not fewer," says Art Haubold, chief of media relations for the NRO. "Today you've got requests for Bosnia, the Middle East, Afghanistan. There are probably more areas today, in a broader sense, than we had to worry about during the Cold War. There are a lot more places that are volatile, now that the Cold War is over. And today we do a lot more work with the war fighter, and those guys have a huge demand for imagery, both strategic and tactical."[20] Yet it is difficult to accept the notion that the NRO has more to keep track of today than it did, say, in the 1970s, when, with a vastly less powerful spying apparatus, it had to keep track of not only Soviet forces but wars and revolutions in Vietnam, the Middle East, Angola, Ethiopia-Somalia, Nicaragua, Iran, and countless other "hot spots."

An agency such as the NRO, of course, which designs and deploys billion-dollar pieces of equipment, works with a long lead time for hardware. Even if the agency had decided to cut back drastically on its future spending as the Cold War waned, it could have taken years, even a decade or more, for those cutbacks to work their way into the system. "When you cut the budget for an organization like this, you're cutting things way down the road," says Haubold. "The satellites that we deploy are expensive and sophisticated, and they are handcrafted. And you've already got satellites in the pipe."

Over the next five years, however, the NRO will begin a fundamental transformation of its space-based spy systems, shifting from the gigantic Advanced KH-11-type satellites to a flock of so-called "small sats," eyes in the sky that cost perhaps a tenth as much and that can be placed into orbit without the large, expensive Titan 4-type launch vehicles. In 1997, Keith Hall, Director of the NRO, speaking about his agency's "future imagery architecture," referred obliquely to the new satellites: "They are smaller, [but] not across the board, because there are some missions that we don't know how to do in a smaller package.... What we have found is that by going smaller in some of our areas, such as our imagery satellites, we can deliver better capabilities taking advantage of newer technologies and do the whole job better and cheaper than with the systems we're currently flying."[21] The small sats will provide certain built-in advantages. Since there will be many more of them than the current four or five large ones, they will presumably provide increased coverage of targeted areas. Smaller satellites will also be harder to detect by those under surveillance. Yet it is by no means clear that the overall cost of the program will fall, since there will be many more small sats (perhaps dozens) to replace the big, lumbering ones.

One possible way to save money is for U.S. intelligence to use commercial imagery rather than depend on the NRO. Those who suggest going commercial certainly know that the NRO's capabilities far exceed anything that exists in the private sector or, for that matter, in the hands of foreign governments. On the other hand, commercial firms can supply imagery of startling quality, including the one-meter-resolution images that were state-of-the-art for the NRO not too long ago. The 1996 House Intelligence Committee report urged that the intelligence community take immediate steps to improve the acquisition and use of commercially available imagery, particularly since privately owned satellites might act as a kind of reserve (or "surge") capability in a crisis.[22] Similarly, the Brown Commission endorsed increased reliance on commercial systems, noting that the private systems would make use of assembly-line and "off-the-shelf" technology, which would mean lower costs.[23] Hall cautiously endorsed recommendations for commercial imagery, saying, "I think our customers really don't care whether the image that they need comes from an NRO system, airplane, or commercial source." The NRO, Hall says, is working with its industry contractors to develop the right mix of government and commercial imagery to meet the needs of the agency's customers, including NIMA.[24] All those customers, of course, are free to shop around for imagery wherever they can find it.

NEW MISSIONS FOR THE TECHINT AGENCIES

Since the end of the Cold War, the U.S. intelligence community has been struggling to chart a new course. Like the CIA, the FBI's national-security division, and the military intelligence agencies, the TECHINT agencies have found themselves adrift, to some degree, in a changing environment. No longer focused exclusively on the Soviet Union and its allies, the NSA and the NRO have faced new demands for tracking drug traffickers and terrorists (targets that John Pike calls "drugs and thugs"), rogue regimes in the third world, and the proliferation of weapons of mass destruction. Many critics of intelligence have questioned whether these new targets require the efforts of a $27 billion spying establishment, yet the reigning metaphor of the 1990s is the assertion of James Woolsey, the former DCI, that even though the Soviet dragon is slain, the world is still plagued by "poisonous snakes" that ought to occupy the attention of the intelligence community.

Certainly, a central mission for the NRO and the NSA is the monitoring of nuclear proliferation. Over the past few years, the TECHINT agencies have closely watched Iraq, Iran, Libya, North Korea, India, Pakistan, and other nations for signs that they are taking steps to acquire or enhance the capacity to build and deliver nuclear weapons. Both agencies report intelligence to the Director of Central Intelligence Nonproliferation Center, established in 1991. According to a report by Robert Walpole, National Intelligence Officer for Strategic and Nuclear Programs,[25] in 1998 the United States monitored medium-range missile tests of Iran's Shahab 3 missile, Pakistan's Ghauri missile, and North Korea's Taepo Dong 1 missile, which "pose an immediate threat to U.S. interests, military forces, and allies," and reported on the development of North Korea's Taepo Dong 2 missile, which could strike Alaska or the Hawaiian islands. (See Chapter 1.) And the TECHINT agencies joined an overall effort to watch Russia, China, and North Korea for signs of assistance to other countries, including rogue states seeking nuclear capabilities. Such monitoring has long been part of the mission of the NRO and NSA.

Yet there are severe limits to the ability of the TECHINT agencies to carry the primary burden for tracking proliferation issues, since governments have become sophisticated in their ability to evade satellite reconnaissance, and much of the threat—for instance, from smugglers who might transfer nuclear weapons from the former Soviet Union to nations or terrorist groups—is not susceptible to overhead surveillance.

A case in point is India's recent nuclear test explosions. According to Toby Dalton, project associate at the Carnegie Endowment for International Peace,[26] India managed to hide its preparations for the tests, in part by feeding disinformation to the United States and in part by conducting its test operations in a manner calculated to mislead the NRO. India was aided in its deception by the fact that the United States had shared information with India about the capabilities of U.S. satellites. For these and other reasons, including an apparent failure by policy specialists and analysts to correctly interpret what data the NRO *did* gather, India's 1998 blasts stunned the world. In the future, the NRO's and NSA's contributions in the area of nonproliferation will continue to be important, but there exist many other mechanisms for monitoring the growth of weapons of mass destruction, including the use of open sources on technology and trade, careful analysis by trained intelligence specialists, diplomatic contacts, the work of the International Atomic Energy Agency, and treaties, such as the new Chemical Weapons Convention.

In the section below, an attempt is made to survey three areas in which the TECHINT agencies have had their work expanded during the 1990s: support to military operations, the environment, and economic espionage. In all three areas the work of the NRO and the NSA, though often productive, raises serious questions about whether such activity can be justified, either in terms of costs and benefits or according to the traditional definition of national security.

Perhaps the biggest new demand for intelligence from the NRO (and to a somewhat lesser extent from the NSA) comes from the U.S. military. From a strategic standpoint, the Department of Defense has always been the NRO's biggest customer, but rapidly advancing technology has given America's TECHINT agencies the ability to supply amazingly detailed, real-time intelligence to field-level commanders and even directly to battlefield infantrymen and fighter-bomber pilots. The NRO calls this "sensor-to-shooter" capability.

The first tactical use of this capability came in Bosnia in September 1996, when F-15E pilots used NRO data supplied to an Air Force/Navy program called the Rapid Targeting Capability. According to an NRO release: "The Rapid Targeting Capability can take video or images from platforms like the unmanned Predator, the U-2, or the Joint Stars and transpose it over an NRO satellite photo and then match the two in real-time to get precise target coordinates.... The process may start with a ten by ten nautical mile picture of the target fed to the cockpit, improved to five by five, then to one by one and finally down to a view that provides the same heading as the approaching aircraft."[27] The release describes the Rapid Targeting Capability as "direct TV to the cockpit." Similar capabilities are emerging for ground forces.

Called SMO in military jargon—for Support to Military Operations— this new technological ability has become increasingly important to the

Pentagon since Desert Storm in 1991. The Pentagon's holy grail regarding SMO is called "Dominant Battlefield Awareness," defined as "the capability to achieve real-time, all-weather continuous surveillance in and over a large geographical area.... The awareness portion of this concept is not limited to enemy activities—it includes awareness of friendly forces, weather, terrain, and the electromagnetic spectrum."[28]

But within the intelligence community, on Capitol Hill, and in policy circles, the military's ravenous appetite for real-time battlefield intelligence raises a concern about priorities. Should the U.S. intelligence system focus chiefly on providing information to diplomats, strategists, and policymakers, or should it be directed to benefit tactical commanders? Both are important, but the fact that the Pentagon controls the budgets of the NRO and the NSA gives the military an advantage in the bureaucratic battles to shape the intelligence system. In recent years, the Defense Department has clearly used its leverage to tilt the balance too far in its own self-interest. (See Chapters 9 and 10.) In the past such questions did not arise, since the means to transfer secure and reliable intelligence directly to the field did not exist. "The debate over SMO is important as it goes to the heart of both requirements and resources," said the 1996 House Intelligence Committee report. "Intelligence is not an easily expanded resource."[29]

The demand for SMO is intensified by the Pentagon's concern over deployments to other countries for peacekeeping and humanitarian efforts, since such operations create demand for useful intelligence about areas of the world outside normal strategic regions, such as Haiti and Somalia, for instance. Concluded the House study: "The current demands being placed on the [intelligence community] to support military operations will make it difficult for the IC to meet the broader national security challenges of the 21st century."[30] It went on to say that "Such goals ... pose daunting challenges for the IC—from both a technological and analytical standpoint—and there are only a few who likely fully understand the ramifications for the IC and for the military.... Some would argue that the 'militarization' of intelligence is already underway...."[31]

Less traditional than the NRO's role in supporting military operations, and much less significant in terms of resources, is the agency's role during the 1990s in monitoring worldwide environmental trends and natural disasters.[32] Beginning in 1992, a team of seventy civilian environmental scientists was paired with a similar number of U.S. intelligence-community officials in an unprecedented task force, whose work sought to answer the question: Can America's Cold War apparatus of high-tech spy gear be mobilized to track a new enemy, the degradation of the environment? The effort was the brainchild of Albert Gore, at the time a senator, who proposed it to Robert Gates, the CIA director. When it began, the effort focused on two distinct areas. The first was backward-looking. In the NRO's archives, compiled over decades, were vast numbers of photographs of 1,600 Soviet missile silos. Perhaps twice a day for all those years, spy satellites snapped photos of these silos, strung out across Central Asia. The NRO's imagery analysts may not have noticed that the snow surrounding the missile silos melted each spring. But to environmental scientists, the existence of a record of snowmelt patterns for a region long closed to scientific inquiry revealed a potential treasure trove of information. Another example: Military intelligence agencies had long kept track of the thickness of sea ice in the Arctic Ocean, in case U.S. missile-bearing submarines needed to surface quickly in a crisis. These and other data, environmentalists believe, could prove to be important indicators of global warming trends.

A second area for the environmental task force was forward-looking. Could the $100 billion-plus array of satellites and undersea devices be used to benefit environmental science? To determine whether the NRO's apparatus could provide useful data, the scientists were allowed access to highly classified capabilities of the KH-11 and Lacrosse satellites. Those capabilities dazzled the scientists. To be sure, environmental science had long been able to use satellites, such as Landsat, which combined the talents of the National Aeronautics and Space Administration, RCA, and Hughes Aircraft to obtain infrared images of land areas and oceans. But Landsat doesn't do close-ups. Where Landsat struggles to achieve resolution of ten meters, resolution as precise as ten inches is a snap for the NRO. Where Landsat shows broad splotches of greenery, the NRO's satellites can count the number of trees in a particular area and even determine their species. Scientists believed that taking a number of highly precise images of small forest areas and using computers to map those points on the broader canvas of Landsat images would yield much more detailed information about remote land areas.

Yet another NRO device—using infrared technology for what the agency calls signature recognition and measurement, or MASINT—was designed to analyze the chemical composition of heat patterns and determine their sources. It could be useful to know, for instance, whether a plume of heat is missile exhaust or a factory emission. Again, the scientific potential is huge: By analyzing the reflection of sunlight off the top of a forest canopy, the NRO's spy satellites can derive a spectral analysis of the composition of the forest, telling scientists whether the forest is deficient in certain chemicals and therefore how healthy it is.

A related intelligence system, run not by the NRO but by the Defense Department, is the Defense Support Program: early-warning satellites designed to detect the flare-up of an ICBM as it emerges from its silo. The Defense Support Program can also be used to look for fires in the Brazilian rain forests, the scientists' task force found.

The scientists' report, which triggered expansion of the environmental task force (since code-named Medea) concluded in part: "Changes in vegetative and desert boundaries, which may be sensitive indicators of global climate change, can be tracked over time by satellite systems. The monitoring of changes in ocean temperature could provide a direct measurement of global warming. Undersea listening systems also may be able to detect this effect by measuring changes in ocean sound speed over long distances."[33]

In a 1996 speech, Director of Central Intelligence Deutch said flatly, "The environment is an important part of the Intelligence Community agenda.... National reconnaissance systems that track the movement of tanks through the desert can, at the same time, track the movement of the desert itself." He added that using the intelligence community's sensors to monitor the environment costs little and does not divert resources from high-priority national-security targets, since the environmental spying "can be done during non-peak hours of satellite use."[34]

Still, the work raises concerns, and not only financial ones. For one thing, it means that the NRO's spy satellites have for the first time been turned against U.S. territory, contravening the long-existing rule that neither the CIA nor any other U.S. intelligence agency may spy on Americans. Though Deutch was quick to add that "to image U.S. territory, we must first get permission,"[35] the work still establishes a troubling precedent. And internationally, the fact that U.S. spy agencies are pledged to track the environment has chilled some of those whose environments are to be tracked. Brazil, for instance, would find events deep in its rain forests suddenly the subject of NRO imaging. At the 1992 Rio de Janeiro summit on the world environment, some third-world nations expressed concern about language in the treaty that seemed to allow the United States to spy on them. According to the World Resources Institute, the nations tried to avoid the word "monitoring" because it sounded too much like surveillance.[36]

Many of these nations undoubtedly remember that, dating back to World War II, the U.S. intelligence community has often disguised its military intelligence collection as part of an environmental monitoring effort. Before the era of spy planes and satellites, the CIA sent a raft of balloons drifting over the USSR to photograph military sites. When the Soviets discovered the balloons, Washington said that they were designed to photograph cloud formations. Later, the earliest U-2 spy-plane squadrons based in London and Turkey were called the 1st and 2nd "Weather

Reconnaissance Squadrons." Perhaps more important, there is a fine line between the collection of innocent data on the environment and strategic information that has a vital bearing on another nation's economic and military security. Information about factory outputs, energy use, agricultural production, and more reveals much about both the environment and a nation's fitness for war.

The fuzzy border between environmental intelligence and economics brings us to the third area that has become an important focus for the U.S. intelligence community: economic espionage. Since the end of the Cold War, the use of America's spy agencies to collect and analyze information on economics, trade, and technology has increasingly been on the agenda. In one sense, that is nothing new: The intelligence community has long been a source of information about economics for U.S. policymakers. The IC has always kept track of economic trends, such as oil and crop production; monitored the economic and trade policies of foreign governments; and pried into other nations' (and companies') use of bribes and subsidies to win contracts and sales abroad. In the area of technology, the IC has identified key industries, such as telecommunications and electronics, as worth watching, since advances in these areas could have an effect in the military arena. For the most part, however, throughout the decades of the Cold War economic intelligence took a backseat to traditional national-security concerns. Only when economics was a component of the global U.S.-USSR rivalry did it become a priority.

But as early as the mid-1980s, under William Casey, the intelligence community began to redirect its attention toward economics. Casey, who had come to the CIA from Wall Street, "believed that economics was going to become more and more a part of CIA's mission, including learning about other countries' economic plans and intentions," according to a former CIA official.[37] Casey began to reengineer the CIA and the intelligence community toward that end, stepping up the hiring of skilled economists, financial specialists, scientists, engineers, and others who could grapple with the issues of trade and technology. Within the CIA, Casey expanded the use of so-called nonofficial-cover officers—that is, spies posing as businessmen overseas, who could more easily travel in business and financial circles.[38]

The intelligence community's increasing focus on economics continued after the Casey era. In June 1990, the CIA created the Office of Resources, Trade and Technology, now called the Office of Transnational Issues. Its job is to track a host of economic issues, including gathering intelligence in advance of trade negotiations and monitoring "environmental trends and civil technology challenges."[39] The CIA also strengthened its National Collection Division, which operates dozens of offices in the United States, changing its name to the National Resources Division, a little-known part of the intelligence community that maintains regular contacts with thousands of U.S. businessmen who travel overseas.

Though most of the economic information gathered by the intelligence community was provided to U.S. policymakers, by the start of the Clinton administration there was speculation that the community could also provide data to assist American companies abroad. At his Senate confirmation hearing as Director of Central Intelligence, James Woolsey described economic espionage, specifically using the intelligence community to aid the competitiveness of U.S. corporations, as the "hottest current topic in intelligence policy." A year later, however, reflecting the Agency's unwillingness to become too entangled in efforts to help the private sector, Woolsey said that the CIA would not share its secrets with U.S. firms and that such a program would be "fraught with legal and foreign policy difficulties." That remains official U.S. policy today.

Woolsey and other intelligence officials since have reaffirmed that although the intelligence community does not provide classified intelligence to American companies for their competitive advantage, it does actively look out for economic espionage conducted by other countries. Indeed, since the early 1990s the FBI's National Security Division, working with the rest of the U.S. intelligence system, has greatly expanded its work in economic counterintelligence, watching both traditional foes and allies for signs that their spy agencies are trying to steal American technology. Few would challenge the need for such vigilance, although the extent of the threat is unclear.

Despite the official U.S. policy against economic espionage, over the past several years there have been reports of U.S. intelligence agencies' involvement in spying tied to trade and technology. In one widely publicized case, several American spies in France and Belgium trying to penetrate France Telecom's telecommunications secrets were caught by French counterintelligence. In another, the NSA intercepted messages between Brazil and France that tipped off U.S. officials to alleged bribery by a French company that was competing with an American firm for a billion-dollar contract in Brazil. The NSA also bugged the conversations of Japanese trade negotiators in Switzerland during critical talks between Japan and the United States over automobile imports.

Oversight

The annual appropriations for the NRO and NSA by Congress are sustained, above all, by a powerful alliance of the intelligence community, its industry contractors, and the congressional overseers who appropriate the funds. A spy satellite drifting silently across suburban Virginia and Maryland would count hundreds of buildings that compose the corporate component of this intelligence-industrial complex. The complex is administered by government officials, but engineered, manufactured, deployed, and maintained by private industry. Around Washington, from Reston and Tyson's Corner, Virginia, to Columbia and Fort Meade, Maryland, the intelligence-industrial complex generates billions of dollars a year in government contracts that go to a handful of big contractors and scores of smaller high-tech firms. The biggest ones, which build and service the NRO's satellites and the NSA's listening posts, can be counted on one's fingers: Lockheed Martin, TRW, Rockwell, Hughes, Boeing, E-Systems, and General Dynamics.

The companies and their lobbyists provide a grateful flow of campaign funds to congressional treasuries. At the same time, they dangle offers of lucrative private-sector jobs for congressional staff and intelligence officials seeking higher pay. The result is a revolving door that allows contractors to cultivate networks of influence. Many senior and mid-level intelligence officials hope to work for the contractors when they leave their agencies, and opportunities for such employment are vast. The analysts who work for private companies that have intelligence contracts exceed by an order of magnitude the number of analysts who work for the CIA. "When I was at the CIA ten years ago, it was understood that if you played ball with the contractors you would get a $250,000 job when you left," said Robert Steele, a former Marine and CIA clandestine-services officer, in 1995.[40]

An outstanding example of the interplay between industry and the intelligence community was the unusual confluence in the mid-1990s that allowed three key members and partners in the intelligence-industrial complex to run it from the inside. They were the Director of Central Intelligence, the Secretary of Defense, and the Under Secretary of Defense for Acquisitions and Technology.

The CIA director, John Deutch, who had taught chemistry at the Massachusetts Institute of Technology, had earlier served on a number of U.S. intelligence advisory boards and on the boards or corporate committees of a wide range of defense and intelligence contractors. These included Martin Marietta, TRW, United Technologies, the MITRE Corporation, and SAIC; overall, Deutch was paid hundreds of thousands of dollars by contractors. It was enough money that upon taking a Pentagon job in 1993, Deutch had to receive a special conflict-of-interest waiver from Secretary of Defense Les Aspin.

William Perry, who became Defense Secretary in 1994, had been involved with industry for several decades. Thirty-five years ago, Perry, a mathematician, built a company called ESL Inc., which was a pioneer in software technology for spy satellites and electronic eavesdropping equipment. In 1978 Perry sold ESL to TRW, which is today a major NRO contractor. Around that time, Perry won for himself the nickname "the godfather of Stealth" for his role in advancing the work on radar-evading technology while he served as chief of the Pentagon's research and engineering work. After Ronald Reagan's election in 1980, Perry stayed on as a top aide to Defense Secretary Caspar Weinberger.

Perry and Deutch, along with Paul Kaminksi, the Defense Under Secretary for Acquisitions and Technology, were part-owners of a small but significant intelligence contractor, Cambridge Research Associates of McLean, Virginia. Though the troika kept a proper official distance from the contracting process, the firm won Pentagon and intelligence-community contracts even while the three men served as the top overseers of the spy world. Cambridge won plaudits for a "virtual reality" system that created living images out of satellite surveillance data for use during the conflict in Bosnia, in conjunction with the NRO.

Though industry executives and intelligence officials work closely together—perhaps too closely—the industry's political-action committees and lobbyists are the ones trying to make sure that Congress keeps the money flowing. For the intelligence community, the keys to Congress are the two intelligence-oversight committees, which bring together a handful of members of each house. These committees, working with a small number of members and staff on the Appropriations Committees in the House and Senate, determine funding levels for intelligence programs, such as the NRO and NSA budgets. Only this group is allowed into the charmed circle that is routinely given information about the supersecret programs of the NRO and NSA. "It's a small community," says a senior lobbyist for a company that is part of the intelligence-industrial complex. "I know every one of the intelligence staffers, and the related appropriations staffers."[41]

Secrecy limits the circle of players on the industry side as well. Lobbying on intelligence programs is so highly classified that in some of the companies' Washington offices, not even the lobbyists themselves have access to the hush-hush nature of the programs. At Rockwell, for instance, a company so closely identified with the intelligence community that the NRO's new headquarters was initially disguised as a Rockwell facility, lobbyists are allowed to usher in specially cleared company officials to meet with congressional oversight staff members, and then must leave the room.[42]

Lobbyists for Lockheed Martin, E-Systems, Rockwell, and TRW agree that because intelligence issues are so complicated, company personnel exert unusual influence with congressional staff. According to Robert

Kohler, a longtime CIA official who later went to work for TRW, direct contacts between contractors and Hill staff have increased in recent years, enhancing the power of lobbyists in the budget process. "There was a time when contractors were not allowed to talk to the committees," says Kohler. "When I was working on this stuff, I would not have allowed one of my contractors in there.... Now, every contractor in the world goes down there and tells his story—and guess what? An awful lot of those stories don't have to do with the national need."[43]

Consider the debate in the mid-1990s over the proposed NRO transition to small satellites. When some members of the Intelligence Committees, with the support of NRO officials, began looking at whether the 30,000-pound KH-11 satellites might be replaced with smaller ones, Lockheed Martin, according to staff members on the oversight committees, lobbied heavily against the switch. The reason? Lockheed had the market sewn up on the big ones, and feared losing the contract if the NRO switched programs. Making the case for Lockheed Martin was Norman Augustine, president of the company and a Washington insider. A confidant of CIA directors and defense secretaries, Augustine was actually mentioned as a possible successor to James Woolsey as director of the CIA. Woolsey, who resigned from the CIA in 1994, was himself a former director of the Martin Marietta Corporation.

But it wasn't just Augustine's cachet that got attention on Capitol Hill. At the time, the home states of the members of the House Permanent Select Committee on Intelligence overlaid with almost artistic perfection a map of the states in which Lockheed Martin had facilities. More than half of the committee, ten out of seventeen HPSCI members in 1995, came from the three states that house Lockheed Martin's largest concentration of plants: California (four members, 27,600 Lockheed Martin employees), Texas (three members, 19,500 workers), and Florida (three members, 15,500 workers). Four of the remaining seven members came from states where Lockheed Martin had a large presence: New Mexico, Colorado, Pennsylvania, and New Jersey.

And then there was Newt Gingrich. Just outside the suburban Atlanta congressional district of the former Speaker of the House is a huge Lockheed Martin plant employing some 12,500 people. Since the earliest days of his career, Gingrich has had a close relationship with Lockheed Martin. At the start of the 104th Congress in January 1995, Gingrich took steps to insure that he would be in a position to defend Lockheed Martin and the intelligence-industrial complex. Breaking with tradition, Gingrich named himself an ex officio member of the House Intelligence Committee, guaranteeing himself and his staff more than a casual acquaintance with intelligence policy.

As the small-satellite debate advanced in 1995, according to committee staff, Gingrich was not a passive player. He often involved himself in the intricate details of the intelligence community's authorization process. Not only that; since the HPSCI is a select committee, its Republican members were handpicked by Gingrich, and they served at his pleasure. Sources close to Gingrich say that he chose members who would favor increased intelligence spending and, by including members from the Appropriations and National Security Committees, helped to grease the wheels under the budget of the spy community. Gingrich himself was a regular recipient of PAC money from Lockheed Martin and other intelligence contractors.

Thus it was not surprising that Norman Augustine's testimony had tremendous clout during the satellite debate. In his public testimony before the HPSCI in late 1995, Augustine declared that proponents of building smaller, lightweight satellite systems would have to wait. "Smallsats certainly represent an intriguing new technology," he said. But "we should continue to work on existing systems so that we will have proven assets to fall back on should smallsats or other new technologies need a lengthy, evolutionary process." Behind the scenes, committee sources said, Augustine and Lockheed pushed even harder, worrying that the switch was too much, too fast, and that the company would suffer financially from the change. In the end, Lockheed Martin's arguments carried the day.

Reforms

Any approach to reforming America's TECHINT agencies has to start with a significant increase in openness. Officials at the NRO and the NSA, no doubt, are reeling from the increased transparency that has already occurred during the 1990s. Yet public scrutiny requires far greater access to the budget, programs, and plans of the TECHINT agencies than currently exists. Most important, for the NRO and NSA to win public confidence, the intelligence community must show that current and future levels of expenditures on high-tech surveillance gear and awesomely skilled manpower are needed to meet specific goals. So far, the intelligence community has not demonstrated, beyond asserting it, that the post–Cold War world requires funding for the NRO and the NSA at or near Cold War levels. (See Chapter 9.)

Second, the U.S. intelligence community must take a close look at whether it can make use of so-called open sources, including commercially available databases and commercial satellite imagery, to replace some of what the TECHINT agencies currently collect. Advocates for the greater use of open sources say that upward of 90 percent of needed intelligence can be garnered through judicious use of readily available data. Certainly, steps have already been made in this direction, and the NRO, in particular, is well aware of the increased usefulness of commercial imagery. No doubt the NRO can provide far more detailed and sophisticated imagery than the private sector—but do the policymakers and the military always need it, or can they make do with lower-quality images in some instances?

Third, a vigorous debate is needed about "mission creep" at the NRO and the NSA. Do the various new uses for SIGINT and IMINT really justify the maintenance or expansion of vastly expensive new intelligence capabilities? Do the armed forces actually require the level of sensor-to-shooter, real-time field intelligence that might be available? How much is enough? The same goes for other new missions, such as environmental and economic intelligence. There, especially, the enormously costly gadgetry of the TECHINT agencies seems unsuitable.

Fourth, the close partnership among the NRO, the NSA, and their respective contractors requires vigilant oversight to protect against featherbedding, sweetheart contracts, and conflicts of interest. Congress ought to enact legislation barring U.S. intelligence officials from going to work for contractors that do business with the officials' former agencies. The same provision ought to apply to congressional staff working on intelligence oversight. And, given the especially sensitive nature of intelligence, contractors' political action committees should be barred from making campaign contributions to members of the oversight committees.

Fifth, the NSA's institutional imperative to be able to decode encrypted messages cannot be allowed to trample on privacy rights. The balance between intelligence and law enforcement on the one hand and the fundamental right to privacy on the other must be carefully weighed.

Notes

1. Estimate provided a decade ago by William Burrows in his seminal book *Deep Black* (New York: Berkley Books, 1988), p. 19. The total today is probably far higher.

2. Estimates by the Federation of American Scientists' Intelligence Resource Program. The FAS estimates the NRO's budget at $6.2 billion, the NSA's at $3.6 billion, and NIMA's at $1.2 billion.

3. See, for instance, the *New York Times*, "Congressional Panels Take Back $1 Billion from Satellite Agency," Sept. 27, 1995.

4. From National Security Council Intelligence Directive 6, cited in Jeffrey Richelson, *The U.S. Intelligence Community* (Boulder: Westview Press, 1995), pp. 25–26.

5. The best source on the NSA is James Bamford's *The Puzzle Palace: A Report on America's Most Secret Agency* (Boston: Houghton Mifflin, 1982). The

Codebreakers by David Kahn (New York: Scribner, 1996) provides a detailed look at cryptography and the NSA's role. Richelson's *The U.S. Intelligence Community* provides a concise summary of the NSA's history and mission. Also useful is a series that appeared in the *Baltimore Sun*, "No Such Agency: America's Fortress of Spies," December 3–15, 1995, by Scott Shane and Tom Bowman.

6. Data on the NSA's Fort Meade facility and staff from the *Baltimore Sun*, December 3–15, 1995.
7. See Richelson, *The U.S. Intelligence Community*, pp. 185–191.
8. Author interview with John Pike, 1998.
9. Norman Polmar and Thomas B. Allen, *The Spy Book: The Encyclopedia of Espionage* (New York: Random House, 1997), p. 389.
10. Author interview with former CIA official, 1996.
11. Polmar and Allen, *The Spy Book*.
12. Federation of American Scientists estimate.
13. For a discussion of technology and its impact on national security, see Whitfield Duffie and Susan Landau, *Privacy on the Line: The Politics of Wiretapping and Encryption* (Cambridge, MA: MIT Press, 1998), pp. 77–108.
14. *New York Times*, July 17, 1998, p. B1.
15. Commission on the Roles and Capabilities of the United States Intelligence Community, *Preparing for the 21st Century: An Appraisal of U.S. Intelligence*. (Washington, DC: Government Printing Office, March 1, 1996), p. 125.
16. *Preparing for the 21st Century*, p. 96.
17. Permanent Select Committee on Intelligence, House of Representatives, *IC 21: The Intelligence Community in the 21st Century* (Washington, DC: Government Printing Office, April 9, 1996), p. 121.
18. *IC 21*, p. 125.
19. Author interview with Pike, 1998.
20. Author interview with Haubold, 1998.
21. NRO FOIA Documents, "Mr. Keith Hall Interview with Aerospace Reporters," April 30, 1997.
22. *IC 21*, p. 142.
23. *IC 21*, p. 120.
24. Hall interview, pp. 15–16.
25. Speech by Walpole to Carnegie Endowment for International Peace, September 17, 1998.
26. Author interview with Dalton, November 1998.
27. NRO release, "NRO Provides Support to the Warfighters," April 28, 1998.
28. Department of Defense, Annual Strategic Intelligence Reviews, cited in *IC 21*, p. 253.
29. *IC 21*, p. 244.
30. *IC 21*, p. 248.
31. *IC 21*, p. 253.
32. See Robert Dreyfuss, "Spying on the Environment," *E: The Environmental Magazine*, February 1995. Also "The Environment on the Intelligence Agenda," speech by John Deutch to the Los Angeles World Affairs Council, July

25, 1996.
33. Copy of preliminary CIA Environmental Task Force report, obtained by author.
34. Deutch speech, July 25, 1996.
35. Deutch speech, July 25, 1996.
36. Author interview with Daniel Turnstall, World Resources Institute, 1994.
37. Author interview with retired CIA official, 1994.
38. See Robert Dreyfuss, "The CIA Crosses Over," *Mother Jones*, January/February 1995.
39. CIA background paper.
40. Author interview with Steele, 1995. Much of the information in this section of the chapter is adapted from Robert Dreyfuss, "Orbit of Influence: Spy Finance and the Black Budget," *The American Prospect*, March–April 1996.
41. Author interview, 1995.
42. Dreyfuss, "Orbit of Influence."
43. Author interview, 1995.

RICHARD A. STUBBING

9 Improving the Output of Intelligence
Priorities, Managerial Changes, and Funding

THE PRIOR CHAPTERS have identified a wide variety of problems hampering the performance of the intelligence community. This chapter addresses current budget allocations, priorities, organizational and managerial procedures, intelligence performance, and possible corrective actions. Finally, it proposes a new mix of inputs—priorities, managerial changes, and funding—to improve U.S. intelligence and better serve the President, top decision-makers, and the nation.

First, a mini-test for the reader:

1. What U.S. national-security support function spent $27 billion to $28 billion in 1996, more than the entire defense budget of all but six nations?
2. What national-security support function has a budget ten times that of the State Department's diplomatic service?
3. What national-security support function is still structured mainly to counter the military threats of the Cold War?
4. What national-security support function apparently missed the most critical events of the Cold War—the long-term economic decline and political collapse of the Soviet Union?
5. What national-security support function doubled the size of its budget in the 1980s while its overall performance arguably declined?
6. What national-security support function spends about $4 billion annually—15 percent of its resources—on an outmoded system of classification and classified documents?
7. What national-security support function gets low ratings from top political and policy officials for accurate and timely forecasts on international events?
8. What national-security support function has had five different leaders in the past ten years, effectively precluding major efforts at innovation and restructuring?
9. What national-security support function has a leader who lacks the managerial and budget authority to go with his overall responsibility?

The correct answer to each question: the U.S. intelligence community.

How did we get to this unacceptable situation? Let's examine the organizational structure, budgetary trends, and performance of the intelligence community since it was formed at the start of the Cold War.

COLD-WAR BUDGETS (1947–1998)

The foreign-intelligence budget has always been classified and is buried within the defense budget. The initial budgets were quite small, but they grew rapidly as the Cold War heated up. More than 90 percent of the intelligence budget pays for collection resources—a profusion of electronic and imaging collection from planes, ships, ground stations, and satellites, along with clandestine human intelligence collection. The other major role of intelligence involves the analysis and production of finished intelligence reports for the national-security team. By the late 1970s, the intelligence programs averaged 5 percent of the total defense budget. President Reagan and Director of Central Intelligence William Casey more than doubled the intelligence effort, which grew to 10 percent of the total defense budget in the 1980s. Since the end of the Cold War, intelligence funding has declined from its peak levels, but with the drop in overall military spending, it still accounts for about 10 percent of the total defense budget.

What makes up the current intelligence budget? Early in 1998, Director of Central Intelligence George Tenet revealed that the total budget for intelligence is about $27 billion; he declined to reveal the major pieces within this total, other than to say that thirteen agencies are involved. We do know that the intelligence budget has three major components:

- The National Foreign Intelligence Program (NFIP) includes national-level intelligence programs and counterintelligence agencies, ostensibly under the DCI. These include the CIA, the NSA, the NRO, the Defense Intelligence Agency (DIA), selected activities of the military services, and some functions of the Departments of State, Justice (the FBI), the Treasury, and Energy.
- The Joint Military Intelligence Program (JMIP) provides intelligence information and support to various consumers within the Defense Department. This includes two major activities: the National Imagery and Mapping Agency (NIMA) and the Defense Airborne Reconnaissance Office.
- The Tactical Intelligence and Related Activities (TIARA). These include defense functions that are essential to tactical operations conducted by our military forces. TIARA programs within each service report primarily to their own operational commanders, and each service's

TIARA program competes for funding with all other combat and combat-support programs.

Data collected from open sources suggest the major pieces of the $27 billion intelligence budget for 1998:

NFIP ($17 billion to $18 billion)
  NRO, more than $6 billion
  NSA, about $4 billion
  CIA, about $3 billion
  DIA, about $1 billion
  Other defense/nondefense resources, about $3 billion
JMIP (more than $2 billion)
  NIMA, more than $1 billion
  Defense Airborne Reconnaissance Office, about $1 billion
TIARA (about $7 billion to $8 billion)
  No details. Includes a multitude of defense activities supporting operational forces.

THE BUDGET PROCESS

The annual budget cycle for intelligence meanders through an eighteen- to twenty-one–month gestation period before funds are finally appropriated by Congress. The cycle, which coincides with that of other government agencies, can be grouped into four phases:

1. Each spring and summer, the intelligence agencies begin to set their priorities and budget requests for the next fiscal year (which commences fifteen to eighteen months later), within tentative budget guidelines from the DCI. These guidelines are often ignored by the individual agencies. For defense intelligence, the more important guideline is that imposed by the Secretary of Defense, since all these agencies and their personnel work for Defense.
2. Each fall, the President's budget is hammered out. Intelligence issues are settled by the DCI, the Secretary of Defense, and the Budget Director, with unresolved questions going to the President for a decision. The President's entire federal budget goes to Congress each February for the fiscal year that begins eight months later.
3. Congressional review takes six to nine months, and includes hearings before the House and Senate Intelligence and Armed Services Committees and testimony before select members of the House and Senate Appropriations Committees. The intelligence committees pass authorizing legislation for intelligence each year. Finally, the intelligence

appropriation bill is approved by Congress and signed by the President, ideally before the October 1 start of the fiscal year. This law permits the agencies to spend funds.
4. Budget execution. The intelligence agencies commit and spend funds approved for the fiscal year in accordance with the guidance received from Congress.

The budget decisions for intelligence tend to follow the traditional bureaucratic approach of least resistance, involving minor, incremental changes in each activity or agency rather than sharp shifts in priorities or funding. Only occasionally does this pattern change. It did so in the early 1980s, when William Casey demanded and received a major infusion of funds for all intelligence activities, with special added funds for CIA covert action, and it did so again in 1998, when the intelligence community received a large increase. As a rule, however, the incremental approach to budgeting permeates the entire budget process for intelligence, including the executive branch and the Congress. This helps explain the patterns of funding for each of the major intelligence activities. During the Cold War, military intelligence on the Soviets and their allies received top priority. Today, military intelligence still receives top priority, even though traditional military threats have sharply diminished and far greater attention should be given to terrorism and the proliferation of weapons of mass destruction. The recent failure to detect preparations for the nuclear tests in India reflects the low military priority assigned to the Asian subcontinent.

The role of the Director of Central Intelligence in this budgetary process is fundamentally flawed and requires remedial action. The DCI has the statutory responsibility for foreign intelligence and the authority to participate fully in the preparation of the budget and in the setting of collection priorities. However, he lacks the ability to impose his priorities on most of the agencies theoretically under him. The DCI has direct control only over the CIA, with its $3 billion budget. The Secretary of Defense ultimately controls the NRO, NIMA, the NSA, the DIA, and all joint and tactical intelligence activities, with total budgets exceeding $24 billion. Similarly, the intelligence-related budgets of the FBI, the DEA, and the Treasury, State, and Energy Departments are under the direct control of each agency, not the DCI. The DCI may advise and suggest priorities for these activities, but the Secretary of Defense and the other agency heads make the final decisions on what will or will not be funded. That basic management principle—the person who pays and promotes controls behavior—is alive and well in the intelligence community. Unfortunately, the current organizational setup contributes significantly to its shortfalls in performance.

What is the effect of this arrangement? Defense decision-makers naturally give top priority to military intelligence. Thus, defense-intelligence agencies dedicate their efforts to satisfying this need, while assigning lower priority to political and economic intelligence. During the Cold War, some 60 percent to 70 percent of total intelligence resources were devoted to the military threats from the communist nations: the Soviet Union, Eastern Europe, and China. Today there is no longer a serious conventional military threat to the United States, yet intelligence still allocates more than 60 percent of its total efforts to military needs. Criticism of national collection assets during the Persian Gulf War further solidified this pattern.

The DCI faces similar limitations in dealing with domestic agencies. The FBI, for example, has statutory responsibility for all counterintelligence efforts in the United States, including ferreting out spies and identifying and tracking terrorist and militant groups. The FBI reportedly spends about $500 million annually on counterintelligence. The DCI has similar responsibilities overseas: tracking terrorist groups and rogue states that sponsor terrorist attacks in the United States and abroad. The responsibilities of the FBI, the DCI, and other agencies—Energy, for instance—clearly overlap, but the record of cooperation is spotty at best. Poor communication and internal management shortcomings hurt performance; the ability of the CIA double agent Aldrich Ames to escape detection for nine years and the long leakage of nuclear secrets to China from the national labs reflect this gap.

The DCI can and does impose his priorities on CIA activities. This was brought home to me in discussions with the CIA Station Chief during an official trip to Japan in 1979. When asked to identify his highest-priority task, the Station Chief promptly identified the recruitment of Soviet or other communist-nation officials who could be turned. This was the top priority in the 1970s for all CIA stations; political and economic intelligence received a much lower ranking. Yet Tokyo provided a window to Asia and fertile opportunities for important political and economic intelligence on China, East Asia, and the Pacific Ocean portions of the Soviet Union. The inability of individual stations to shape their mission to local conditions proved a surprise. Perhaps a partial explanation lay in the fact that neither the Station Chief nor his deputy spoke Japanese or had had any exposure to the Japanese culture before arriving in Tokyo.

Budget Decision-Making

Establishing overall intelligence funding levels and priorities has not been the forte of the DCI, the White House, or the Congress. Even the prestigious 1996 Brown Commission avoided the question of funding guidelines for intelligence. Rather, the minutiae of many small budget issues and the dominant role of the Secretary of Defense overwhelm the budget process. This situation also reflects the operating imperative of agencies to continue doing whatever they have done in the past. They seldom take critical looks at such fundamental questions as the overall level of intelligence; the proper blend of imaging, electronic, and human intelligence collection; and needed improvements in intelligence analysis and output.

A revealing example of the process occurred in the late 1970s, when Admiral Bobby Inman was director of the NSA. Each fall, the Office of Management and Budget examiner, equipped with the proper clearances, spent several weeks at NSA headquarters reviewing its proposed budget in great detail. For three successive years, this examiner identified a large sum of new funds requested under the mundane title of "building modernization" or "upkeep of the facility." The agency officials could provide no justification for these funds, and the examiner recommended their deletion from the budget. At precisely this point each year, Admiral Inman requested and was granted a private meeting with the Budget Director or the Associate Director of the OMB. No examiners were present. At this meeting, Inman revealed that the added funds were for some hush-hush development in electronics or cryptology that could provide a major "breakthrough" in the Cold War. The need for secrecy forced Admiral Inman to mislead the OMB staff. Not surprisingly, each year the Budget Director approved the funds. The lesson: Dissembling in the name of national security has its rewards. Did these projects ever make a big difference? Not to the knowledge of the OMB, but the information was so secret that one could never be sure.

Even when the dollar level of intelligence budget issues is substantial, the budget imperative of continuing past practices prevails. President Carter called a meeting of his national-security advisers in December 1978 to resolve a budget debate over a new, $5 billion imaging satellite for the NRO. The chief advantage of the new satellite was its ability to produce sharper pictures with higher resolution than the satellite then being used. The NRO argued that these improvements would provide a high payoff in improved intelligence on Soviet forces and weaponry. The OMB's analysis, on the other hand, concluded there would be little added value from this new program; the operational satellite already had a high degree of resolution that met almost all our needs. The OMB proposed continuing the existing satellite program, with savings of several billion dollars.

The meeting took place in the Cabinet Room of the White House on a brisk December morning. The small group in attendance included the President, Secretary of Defense Harold Brown, Director of Central Intelligence Stansfield Turner, and Associate Director Randy Jayne of the OMB. President Carter opened the meeting by stating that he had read all the background papers on the issue and had one question. Then, instead of asking about the proposed satellite, the President asked about the design of the engines on the third stage of the booster. A stunned silence followed; the booster engines had nothing to do with the issue (the operational satellite and the new satellite would use the same booster to reach orbit), and the principals at the table had no clue. Finally, an Air Force brigadier general on assignment to the NSC staff responded, and the balance of the forty-five–minute decision meeting was a rambling technical discussion on booster-engine design between the President and the officer. The meeting ended with no serious discussion of the new imaging satellite and no decision on the issue. Several days later, all parties were informed that the President had approved the new satellite program. (In retrospect, he had probably decided even before the meeting.) The lesson: Technological advances and the political attraction of new business for contractors win out most of the time over issues of budgetary restraint and marginal added benefits.

Intelligence Performance: The Cold War and Beyond

Intelligence proved consistently successful in reporting on the status of Warsaw Pact military forces and their weaponry, and in tracking military conflicts, such as the Iran-Iraq war and the conflict in Afghanistan. However, the record is far more modest in intelligence's most important role: giving the President and his advisers warning of impending foreign-policy crises. Significantly, the driving priority for the future must be more accurate information in exactly this area of subpar performance. Our intelligence resources are not properly positioned for the twenty-first century.

Let's look at the record for supporting evidence. First, two disclaimers: Some intelligence successes, by their very nature, help prevent crises and are therefore invisible and unreported. Also, good intelligence data have been available on many important events; the inability of the total system to provide advance warning does not negate this valuable work. In what follows, I will confine this analysis to intelligence reporting rather than covert action, which is covered in Chapter 2.

The military capabilities of the Soviet Union and its allies had the highest priority during the Cold War. Multiple collection systems involving satellites, ground-based systems, and human agents produced accurate static reports on hostile military forces. Predicting the actual use of these forces proved far more difficult. Aerial reconnaissance successfully tracked the Soviet deployment of missiles to Cuba in 1962, overcoming an earlier estimate that the Soviets would not deploy them. Intelligence failed to predict the Soviet invasions of Hungary in 1956, Czechoslovakia in

1968, and Afghanistan in 1979. Overall projections of Soviet military capability and the characteristics of their weapons systems were consistently overstated. Probably the single most important shortcoming was the decades-long failure to identify the true size and growth rate of the Soviet economy, with major implications for U.S. foreign policy. Finally, as Melvin Goodman shows in Chapter 2, the consistent inability to foresee the collapse of the Soviet empire was a serious failing.

In many ways, China remains an enigma to our intelligence community. Again, there has been good reporting on forces but no early warning of the Chinese invasions of Korea, Tibet, and Vietnam. Political and economic forecasting has proved an extremely tough nut, with few successes. The community also failed to predict the military intervention in Tiananmen Square.

In East Asia, there was no warning of the North Korean invasion in 1950. Military intelligence on the progress of the Vietnam conflict was consistently optimistic and proved erroneous; CIA estimates were more realistic but also often erred on the rosy side. In South Asia, U.S. intelligence did not detect in advance the Indian decision to test a nuclear device in 1974, and, as we have seen, failed again in 1998.

The intelligence record on events in the Middle East is unsatisfactory. One important success: correctly forecasting the 1967 Israel-Egypt war. Unfortunately, warning was not provided for the 1973 "Yom Kippur" war, the 1982 Israeli invasion of Lebanon, or the 1983 terrorist bombing that killed 250 U.S. Marines in Beirut. Events in Iran and Iraq proved impenetrable. Intelligence missed the Iranian revolution and the fall of the Shah in 1979, Iraq's invasion of Iran in 1980, and the decision by Iraq to invade Kuwait in 1990. Estimates of Iraqi military capability in the 1991 Persian Gulf conflict were greatly overstated.

In the 1990s, as intelligence made a transition from the Cold War, the record continued to be spotty at best. General H. Norman Schwarzkopf's harsh criticism of national intelligence input during the Persian Gulf War led to even tighter control of satellite imaging resources by the military, contributing to the creation in 1996 of a single imaging agency under the control of the Defense Department. This agency helped pinpoint the new nuclear-weapons facility in North Korea, but also reflected the low military priority assigned to India. Intelligence reporting proved uneven in assessing the situations in Somalia (1992–93) and Haiti (1994). There was no specific warning of the 1996 terrorist attack on the Khobar barracks in Saudi Arabia, nor of the 1998 terrorist bombings against the U.S. embassies in Tanzania and Kenya. On the economic front, the intelligence community neither foresaw the recent economic downturn in East Asia nor accurately tracked subsequent developments.

The politicizing of reports has been a serious challenge to accurate intelligence. The Pentagon's consistent overestimates of Soviet weapons capability suggest that inflated threat estimates serve as justifications for new weapons and larger defense budgets. During the Nixon administration, defense intelligence erroneously concluded that the large Soviet SS-9 strategic missile had multiple–nuclear-warhead capability. The CIA dissented from this view, leading to a major confrontation with Defense (backed by Henry Kissinger). Defense attributed exaggerated capabilities to the Soviet Foxbat fighter aircraft in the early 1970s; these estimates helped to justify the Air Force F15 Eagle fighter. In 1980, congressional reluctance to approve production of the Army's M-1 Abrams tank evaporated after an intelligence briefing described a projected new Soviet tank with frightening capabilities. The threat, however, was a complete myth, drawn from the imaginations of Army officials. Naval intelligence in the late 1970s expressed great angst over the first observed Soviet aircraft carrier. Not mentioned was that this new "carrier" was far smaller than U.S. carriers, lacked catapults, and could not accommodate first-class fighter aircraft. The overblown threat helped to gain congressional approval for a large, new U.S. aircraft carrier.

In the 1980s, William Casey often imposed his predetermined views of the worldwide communist menace on intelligence reports. His inaccurate and embarrassing 1985 reports on hostile Soviet military intentions toward Iran and the identification of "moderates" in the Khomeini regime overruled the contrary findings of his own analysts at the CIA. In a chilling critique, Secretary of State George Shultz told Deputy Director Robert Gates: "I wouldn't trust anything you guys said on Iran no matter what. You have a very dissatisfied customer. If this were a business, I'd find myself another supplier." In the assassination attempt on Pope John Paul II, Casey directed the CIA analysts to make the best possible case for Soviet involvement despite the clear preponderance of evidence to the contrary.

What lessons were learned about intelligence performance during the Cold War? Certainly, the emphasis on Soviet and Chinese military forces over worldwide reporting contributed to the weak overall record. We learned much about the makeup of the Soviet military, but were unable to predict Soviet incursions into neighboring countries or to estimate the real state of the Soviet economy. Nor did the intelligence community predict many crises around the world. Finally, covert actions using paramilitary forces or other means to overturn unfriendly governments had only slight success, generated much negative publicity, and often left lasting negative consequences.

Post–Cold War Intelligence Priorities

The nation needs a revised agenda for the intelligence community as it confronts a different geopolitical scene. Top priority must go to impending crises in developing nations, to the growing threat from terrorists, and to the proliferation of weapons of mass destruction. Lower priority now must go to military intelligence, reflecting U.S. military superiority. Mission objectives need to be realigned and resources redirected to support these new priorities. The DCI's role as manager of intelligence resources needs to be strengthened, and a knowledgeable, independent-thinking person must be selected to lead the intelligence community after the November 2000 election.

This section will examine three major tasks in developing needed reforms: to realign mission priorities, to change the mix of intelligence resources, and to strengthen the role of the DCI.

Realign Mission Priorities

During the Cold War, the top priority in the allocation of intelligence resources was information on the nuclear and conventional military forces of the Soviet Union, its Eastern European allies, and China. Today, the old threats have greatly diminished. The Soviet empire has dissolved, and communism is no longer viewed as a serious threat. Eastern Europe now has close ties to NATO and the West. China has a growing market economy, and the United States and its allies are among its leading trade partners. The Russian military still possesses a strong arsenal of nuclear weapons, but these are being downsized following treaties with the United States. Russian conventional forces have been reduced to a fraction of their former size and pose no threat to move beyond their borders. The Russian economy is struggling in its transition to the free market. The closed Soviet society is now open; China, too, is far more open than in the past. Military intelligence requirements for watching these former communist nations have dropped sharply.

The United States is the single remaining military superpower. The International Institute for Strategic Studies identifies about $800 billion in worldwide defense expenditures in 1997. U.S. defense spending alone accounts for $270 billion—a third of the total. The United States and its allies in Europe, Asia, and the Middle East together spent more than $600 billion, almost 80 percent of the worldwide total. By contrast, Russia spent $65 billion and China $35 billion, while the rogue states of North Korea, Iran, Iraq, Syria, and Libya spent a total of $14 billion. The U.S. defense effort is thirty times that of North Korea and Iraq combined, seven times that of China, and four times that of Russia.

The United States and Russia have ceased production of nuclear warheads and are reducing their nuclear arsenals. START II calls for both nations to reduce their strategic nuclear-weapon inventories to one-third of Cold War levels (though the treaty has not yet been ratified by the Russian Duma, as of summer 1999). Tactical nuclear inventories have dropped even more sharply. China, France, the United Kingdom, and Israel are the other major nuclear nations. India and Pakistan are new members of the nuclear club, but each has only a handful of weapons. Monitoring the implementation of nuclear treaties remains an important role for intelligence, but the overall level of reporting required for nuclear weapons is far less than during the Cold War.

As Roger Hilsman points out in Chapter 1, the demands of military intelligence on conventional forces have also shifted dramatically. Though the probability is remote, attacks against our allies in volatile areas such as the Middle East or East Asia could lead to U.S. military intervention. Demands are growing for intelligence on nations facing civil and military unrest, such as Bosnia and Kosovo, where the United States has intervened in both peacemaking and peacekeeping roles. The new threat to world stability and a prime concern for the intelligence community is the combination of terrorism and the proliferation of weapons of mass destruction—nuclear, biological, and chemical.

Political reporting on impending foreign crises needs to improve over the anemic track record of the past thirty years. Recent intelligence failures in Somalia, Haiti, India, Saudi Arabia, Tanzania, and Kenya are an unacceptable standard for the future. Ongoing events in developing nations need to receive greater attention in the assigning of intelligence resources, although, as already pointed out, most of these needs could be met by enhancing the resources of the U.S. diplomatic service in the field.

Economic reporting on important happenings in such key areas as Russia, China, East Asia, and the Middle East is also important, although much of this information can be gathered from open sources. Environmental problems are a related and growing concern; policymakers need more accurate information on the types and causes of pollutants worldwide. These needs may be met through greater use of commercially available resources. Our national intelligence resources do not and should not conduct economic espionage against foreign companies; this is not a proper role for U.S. intelligence.

The preceding argues for a substantial realignment in national intelligence targeting and reporting. Top priority should go to terrorist threats and the proliferation of weapons of mass destruction. Lower priority should go to traditional military intelligence. How significant are these shifts in focus? Today, about 60 percent to 70 percent of national intelligence resources track military targets; the rest follow nonmilitary activities. As a rough guideline for the future, the balance should be reversed. Nonmilitary targeting should grow to account for 60 percent to 70 percent of the intelligence effort, with the remainder allocated to military needs.

Change the Mix of Intelligence Resources

The substantial shift in missions and targeting priorities will change the mix of intelligence resources in the next five years. For openers, a top-down review of requirements must replace the incremental budgeting approach used in the past. Collection assets, in particular, should be downsized and refocused. More resources should go to the U.S. diplomatic service and less to CIA field operatives. The latter should focus on a few priorities: rogue or threatening states, terrorism, and the proliferation of nuclear and biological weapons. On the other side, the mediocre record of foreign intelligence reporting in the past also suggests the need to expand and strengthen the intelligence community's analytic capability, including the use of outside resources, particularly from the academic world.

Conversely, the lower priority assigned to military reporting places less dependence on sophisticated satellite and ground systems. The use of covert actions—a vestige of the Cold War—could be terminated as a tool of U.S. policy to promote or subvert foreign governments. Reduced emphasis on classification and secrecy also seems in order. Finally, this combination of actions will permit a decrease in the level of intelligence funding.

About $27 billion to $28 billion is spent annually on intelligence. Of this, the National Foreign Intelligence Program (NFIP), the Joint Military Intelligence Program (JMIP), and the FBI counterintelligence effort receive approximately $20 billion. The remaining money goes for tactical intelligence in support of military missions.

The *NRO* currently spends more than $6 billion annually on imaging collection. The bulk of these costs are for expensive satellites whose primary Cold War mission was to monitor Soviet and Chinese nuclear and conventional forces. Military intelligence reporting on these former enemies can be sharply reduced. Other military needs can be satisfied with less exotic satellite payloads. A reduced level of $4 billion to $4.5 billion annually should provide satisfactory coverage.

The *NSA's* annual budget of $4 billion provides extensive voice and data intercepts and allows for the conduct of cryptographic operations. Much of the need for this information was driven by the closed communist societies of the Cold War. Today we face the opposite problem: information overload. Formerly closed societies are now open, and there has been an explosion of available information from open sources. The massive collection capabilities of the NSA satellite and ground assets multiply this base of information; at the same time, technological advances in fiber-optic cables cut down the effectiveness of satellite collection. Selective reductions are possible in NSA funding, particularly in the area of satellite collection; the agency's budget, currently at $4 billion, could be reduced to somewhere between $3.2 billion and $3.5 billion.

The *DIA* has a $1 billion annual budget and is the principal producer of military intelligence reports for the Joint Chiefs of Staff and the Secretary of Defense. Its functions parallel those of the CIA in clandestine human collection and the production of intelligence analysis, and its work at times overlaps the intelligence efforts of the military services. Consistent with the reduced priority for military intelligence on weaponry and order-of-battle estimates, the DIA level could be reduced to $0.8 billion.

Other national intelligence assets include about $3 billion in *other defense and nondefense resources*. Military assets make up the bulk of this. The deemphasis on military targeting suggests that the resources in this category could be sharply decreased, down to a level of $2 billion to $2.4 billion annually.

The *imaging and mapping* (JMIP) budget exceeds $2 billion annually and includes two major imagery programs primarily for military customers. The decreased requirement for military targeting lessens the need for imagery and for airborne reconnaissance, and provides opportunities for savings of 20 percent, to a level of about $1.6 billion.

The *Foreign Service* is not officially part of the intelligence budget, but it should be a prime source for satisfying the higher priority assigned to political and economic reporting. Yet this once highly capable source of intelligence has atrophied in recent years, due to a combination of shrinking budgets and inadequate use by the Secretary of State. These trends need to be reversed. A more visible and better-staffed presence around the world should lead to higher-quality intelligence reporting. An increase of 10 percent in the budget of the diplomatic service, to a level of $2.5 billion, should suffice.

The *CIA* is a producer of intelligence analysis and the primary collector of clandestine human intelligence. The demand for accurate intelligence on threatening countries, the spreading danger from terrorist groups, and the need to cope with the proliferation of weapons of mass destruction argue for the prioritization of CIA targets. Also, an upgraded analysis capability is required to improve the quality of intelligence reports. On the other side, the CIA should drastically reduce its presence in friendly and neutral states, leaving intelligence collection generally to the diplomatic service. Implementing these actions will permit a smaller

CIA; a decrease in the CIA budget of 10 percent, to $2.7 billion, would reflect this change.

The conduct of covert activities was a major CIA effort during the Cold War. These actions, involving the use of U.S.-funded paramilitary forces and other unsavory actions to overthrow unfriendly governments or to retain friendly ones in power, simply do not coincide with the democratic principles in which we believe. Covert activities for these purposes should be discarded as a part of U.S. foreign policy.

The *FBI* is responsible for counterintelligence efforts within the United States and in our embassies overseas. Funding for the intelligence division of the FBI has grown rapidly in recent years, to an estimated $0.5 billion annually. The growing domestic threat from terrorist and militant groups (possibly armed with nuclear, biological, or chemical weapons), and the vulnerability of national information systems to sabotage, support these increases. The CIA has similar responsibilities for counterterrorist activities in foreign countries and should expand these efforts. The often frayed relations between these two agencies cannot continue. The DCI and the FBI Director need to strengthen their working relationship to improve the effectiveness of counterintelligence. Expanded FBI antiterrorism efforts could require a substantial budget increase of $0.1 billion to $0.2 billion.

Published estimates suggest that 15 percent of the intelligence budget, some $4 billion annually, supports the *classification and compartmentalization* of intelligence data. Now, as suggested by Kate Doyle, is an excellent opportunity to declassify much of this old material and to downsize future classification efforts. Many official historical reports on past U.S. foreign-policy events are incomplete, lacking any CIA input. Critics describe the current procedures as an obsession with secrecy and turf protection that hurts the effectiveness of intelligence reporting. A sizable reduction of up to $1 billion annually in the classification effort seems feasible. This could actually improve the performance of U.S. intelligence by eliminating roadblocks to the sharing of critical information. These cuts would be included in the reductions of the various intelligence agencies' budgets.

Organizational and Managerial Needs

The Director of Central Intelligence operates today in an organizational nightmare. As I have noted, more than 85 percent of the foreign-intelligence budget is controlled and executed by agencies (primarily Defense) not under his control. Numerous executive orders and the 1993 Intelligence Authorization Act give the DCI the authority on paper to establish requirements and priorities for the entire intelligence community, but these cannot be implemented under the current structure.

What should be done? To improve the effectiveness of intelligence and to meet the shifting priorities that place greater emphasis on nonmilitary reporting, the DCI must have greater control in peacetime over the NRO, the NSA, and NIMA. Organizationally, these three agencies can still remain in Defense, but in peacetime the DCI should have the authority to hire the directors of these agencies after consulting with the Secretary of Defense. The DCI would also have primacy in setting the operational assignments for these three agencies, again after consulting with the Defense Secretary. The DCI should also have primacy in setting agency funding. In wartime, control of all three agencies would revert to the Secretary of Defense.

Bold leadership will be required from the new DCI after the November 2000 election. Without this leadership and the public support of the new President, these proposed reforms will not be implemented. The DCI will be the critical force in overcoming the entrenched bureaucratic ways of the intelligence community and making it a more responsive and effective organization. This person should be knowledgeable, highly respected, and visionary; willing and able to set a new course; and capable of withstanding large doses of criticism from protectors of the status quo. For these reasons, the choice of DCI must not be subject to veto or undue influence by the intelligence community. Solid political skills will be required to gain the support of Congress for reforms opposed by Defense. An effective DCI should have the stature and integrity to give the President frank assessments on intelligence matters, uncolored by what the President wishes to hear. Ultimately, this person must accept the risk of being fired for his candor.

How will the DCI set the priorities for the intelligence community? One possible vehicle is the new NSC Committee on Transnational Threats, proposed by the Brown Commission and established in 1996 to address threats from international terrorism, weapons proliferation, drug trafficking, and organized crime. This committee will establish priorities and insure coordination among the national-security and domestic agencies dealing with these problems. The Brown Commission also recognized the importance of consumer input to the process. It proposed a Consumers Committee, with representation from all the principal producers and users of intelligence reports. This group would report regularly to the Committee on Transnational Threats on how well the needs of intelligence consumers are being met, identifying gaps and shortcomings. This is a useful proposal that should be implemented.

Reforms in the CIA's analytic capability are needed. The touchstone of the intelligence community's performance is its ability to provide accurate and timely reports to its customers. Ever-increasing layers of bureau-

Improving the Output of Intelligence 187

cracy (compartmentalization of information, for instance) and political intervention in the production of reports have contributed to the current unsatisfactory situation.

The first steps should be to expand the number of analysts and to modernize the processing capability to cope with the explosion of both open and classified information. The best minds should be assigned to the production of reports, and competitive analysis should be encouraged. Analysts should travel more frequently to the regions for which they are responsible, to enhance their understanding of the countries and their cultures. The National Intelligence Council of the NSC should expand the use of outside experts in the preparation of national intelligence estimates. Current procedures require CIA analysts to pass their finished intelligence estimates through a gantlet of managerial review levels before publication; this process needs to be refined. Analysts need incentives to take risks in their reports and to defend their estimates before top officials. The process should encourage the best reports, uncolored by political pressures.

GAME PLAN FOR 2001

The preceding section calls for some drastic reforms in the intelligence community to improve its effectiveness against a sharply altered array of threats. This section summarizes what needs to be done and suggests how to obtain the political approval needed to implement this plan after the election of a new President.

Table 9-1 reflects the new priorities that can be phased in by the new administration over the next five years, starting in 2001.

Proposed Organizational and Managerial Changes

In peacetime, the DCI should have the authority, in consultation with the Secretary of Defense, to appoint the directors of the NRO, the NSA, and NIMA, to control their operational assignments, and to set their funding levels. In wartime, control over these agencies would revert to the Secretary of Defense.

Use the newly established Committee on Transnational Threats to provide guidance on intelligence requirements and priorities for collection and analysis. Also establish a Consumers Committee of the NSC with representation from all the principal producers and users of intelligence reports.

Upgrade the CIA's analytic capability by expanding its size, modernizing its processing capability, using more outside experts, protecting analysts from political interference, and creating incentives for bold assessments. At the same time, reduce the CIA's presence in the field, focusing

Table 9-1. Suggested five-year funding changes (in 1998 billions of dollars)*

	1998†	Adjustment by 2005
NFIP (national intelligence)		
National Reconnaissance Office	6.0+	Reduce 1.5–2.0
National Security Agency	4.0	Reduce 0.5–0.8
Central Intelligence Agency	3.0	Reduce 0.3
Defense Intelligence Agency	1.0	Reduce 0.2
Other national intelligence assets	3.0	Reduce 0.6–1.0
JMIP (imaging and mapping)	2.0+	Reduce 0.4
TIARA (tactical intelligence)	7.0–8.0	Reduce 0.7–1.2 (DOD responsibility)
(Classification Costs)	(Est 4.0)	(Reduce 1.0, included in agency cuts)
FBI Counterintelligence	0.5	Increase 0.1–0.2
Total Intelligence	27.0–28.0	Reduce 4.1–5.7
State Department, diplomatic service	2.3	Increase 0.2

*These proposals reflect a reduced emphasis on high-technology collection platforms, a greater emphasis on open-source collection, and a shift away from traditional military targets and toward terrorist groups, rogue states, and the proliferation of weapons of mass destruction.
†Excludes one-time supplement for 1998

it on military threats, terrorism, and the proliferation of nuclear, chemical, and biological weapons.

Covert action to overthrow other governments should cease as a tool of U.S. foreign policy.

How Can This Plan Be Implemented?

The proposals in this plan involve substantial changes in intelligence priorities, management, and organizational structure. The Defense Department and its allies in Congress, particularly the Armed Services Committees, can be expected to oppose many of these proposals, which they will see as a diminution of their role in intelligence. The intelligence contractors will mount a major attack against this plan to avoid losing business. With this level of opposition, what will it take to sell this plan?

The best chance to get these proposals adopted will involve a concerted effort during the "honeymoon" period after the new President takes office in 2001. The President, the Intelligence Committees, and the top congressional leaders of both parties need to be part of the process; these reforms must be seen as bipartisan in nature or they will surely fail. Both parties must agree on the need for substantial changes to improve the effectiveness of U.S. intelligence. The political "game plan" to sell these proposals should include the following:

Improving the Output of Intelligence 189

- During the election campaign, the presidential candidates need to commit themselves to improving the performance of the intelligence community.
- The President-elect needs to select a DCI of high integrity, dedicated to reform, with considerable political skills and some background in intelligence matters. The person selected to be Secretary of Defense also needs to be open to these reforms.
- Following confirmation, the DCI should meet with the chairmen and ranking minority members of the Senate and House Intelligence Committees for a mutual discussion of possible reform initiatives.
- The new President should meet with the Secretary of Defense and the Joint Chiefs of Staff to discuss his commitment to intelligence reform. The President should also establish a Presidential Commission on Future Threats to assess potential dangers; recognized experts from a variety of backgrounds should be appointed. The commission should move quickly and issue its report in the spring of 2001.
- The NSC should conduct a study of intelligence reform in the early months of 2001. The study group should include representatives from the Defense Department, the State Department, the DCI, and the FBI. Several options should be prepared for consideration by the full NSC and the President. Input from the House and Senate Intelligence Committees should be reflected in the options developed.
- The President should decide on a specific package of reforms. He should meet with the Senate and House leadership of both parties to seek their support and to express his willingness to use some of his political capital to achieve the needed improvements. He should strive to obtain legislative approval of the reform package before Congress adjourns in the fall of 2001.
- The President should use the "bully pulpit" to announce the proposed reforms to the American people and explain why they are needed. Gaining the support of citizens will help offset the adverse publicity from agencies and contractors opposed to the plan.

PAT M. HOLT

10 Who's Watching the Store?
Executive-Branch and Congressional Surveillance

THE BASIC DILEMMA with which this chapter is concerned is how an open, democratic society such as the United States can establish public control of activities that are necessarily secret.

The intent of the National Security Act of 1947 was that this control would be exercised by the President. (The act gives the responsibility to the National Security Council, and names the members as the President, the Vice President, and the Secretaries of State and Defense. In practical terms this means the President; he is the only one on the NSC who counts.)

Experience showed that this was inadequate. Some Presidents delegated too much authority and did not pay enough attention to how it was used. Others abused the resources of the intelligence community to carry out, in secret, policies that would have embroiled them in unwanted political controversy if attempted publicly.

In time, Congress bestirred itself to oversee intelligence, just as it does every other activity of the federal government. It established Intelligence Committees in the House and Senate thirty years after it created the CIA. These, too, have proved inadequate. Some of the reasons for this are built into the system; others could be overcome, given the political will to do so.

The President and Congress take different approaches to controlling the intelligence community. The President—that is to say, the national-security establishment, mainly State, Defense, and the NSC—is concerned with two things. One is intelligence. What is happening in foreign countries and what is likely to happen? This involves collection, essentially fact-finding, and analysis, interpreting the facts that have been found. The other is using the varied resources of the intelligence community to advance the interests of the United States (as the President sees them) in foreign countries. This is covert action: secret attempts to assert influence abroad in ways that will not be attributed to the United States.

Congress, on the other hand, devotes most of its attention to avoiding intelligence mistakes that can embarrass the United States, or worse. This focus is largely the consequence of the events that drove Congress to get serious about oversight in the first place. Most of these events involved ill-advised covert actions; some had to do with risky techniques of collection. As Congress has gained experience with oversight and familiarity with the intelligence community, it has given more attention to improving collection and analysis than was the case in the 1970s. But congressional concern is still weighted toward preventing mistakes.

Congress also reacts to what is in the day's headlines and the evening newscasts. This is natural and desirable (though sometimes it is overdone). Congress is a political body, and especially with respect to secret intelligence, it is the public's trustee, attempting to insure that the public interest is being served.

THE PRESIDENT

There have been ten Presidents since World War II, the period during which the United States has been in the intelligence business in a big way. Each has approached the job differently. Some have been more interested than others in the intelligence community, and have therefore paid more attention to what it was doing. One—Bush—had been Director of Central Intelligence. Some Presidents, being less interested, have given the community freer rein and have been content to allow their appointees in the larger national-security apparatus to keep it in check, or try to.

This larger national-security apparatus basically comprises the Departments of State and Defense and the staff of the National Security Council. At one time or another, there have been designed a multitude of interagency committees, working groups, and other bureaucratic devices to consider and approve, amend, or disapprove proposals for intelligence activities. None of these mechanisms has been totally successful in avoiding embarrassment (or worse) precipitated by intelligence failures. But avoiding embarrassment is only a part of oversight. Even if an action fails, there is no embarrassment unless the failure is publicized. A more important question: "Is this a good thing to do even if it works and we can keep it secret?" Or, with respect to collection, "Is the intelligence likely to be gained by this operation worth the costs of failure?"

One reason the bureaucratic mechanisms have not worked better is that the intelligence community has grown and agencies within the community have proliferated. So has the larger national-security community, notably the NSC staff. This unavoidably increases general bureaucratic clumsiness and the difficulty of interagency coordination. It also increases petty interagency rivalries, jealousies, and jurisdictional squabbles. These are problems of size, and they are not unique to the government or to intelligence agencies. They are found also in the private sector. Only a strong, forceful President can deal with this situation.

Another mechanism of presidential control is the President's Foreign Intelligence Advisory Board (PFIAB), which was created under another name by President Eisenhower and refurbished by President Kennedy after the Bay of Pigs. It has had a checkered history, but one that is hard to document because its reports to the President are secret. Some members have been willing to talk about its work; some have not. Some have been political appointees with no knowledge of the intelligence community. Others have had intelligence experience, notably Admiral Bobby Inman, who came to the board as a former director of the National Security Agency. President Carter abolished the board; President Reagan revived it. During long periods of the Clinton administration, the chairmanship has been vacant. The PFIAB has the potential to be a significant instrument of presidential control, but it will have to be substantially strengthened. It needs full-time, knowledgeable members and a competent staff. It also needs effective presidential support in fighting off bureaucratic jealousies and in gaining unfettered access to intelligence secrets.

As a consequence of the growth of the bureaucracy, the President, in a sense, has become its prisoner. In theory, the President controls the bureaucracy through his appointees to the Cabinet and to lesser jobs. These appointees are the President's men and women. Their loyalties are to the President and to his program. But the agencies they are supposed to supervise are run almost entirely by career officials, most of whom are competent, dedicated public servants. Almost all of them are a part of the culture of the agency in question, and they reflect, perhaps subconsciously, its institutional view. Subtly, over time, the President's appointees take on the coloration of the agencies they are supposed to supervise. Perhaps without realizing it, they become as much (or sometimes more) the agencies' advocates to the President as they are the President's surrogates in the agencies.

Finally, White House oversight of the intelligence community has been hampered by the doctrine of plausible deniability. For example, the failure of the U-2 was compounded when President Eisenhower made an initial attempt to deny his role in the affair and was then revealed to have known about it from the beginning. Plausible denial puts the President in an impossible situation. Even when it is successful, it means that the President has wriggled out of a bad foreign situation at the cost of falling into a worse one at home. Inherent in his denial of knowledge of an intelligence operation gone sour is his admission that he was not paying attention to what his subordinates were doing—in other words, that he was not minding the store. In the view of many American voters, that is worse than getting into trouble abroad.

The most recent major use of plausible denial was in connection with the Iran-Contra scandal in the 1980s. This was the effort by the Reagan administration to bypass congressional strictures on aid to Nicaraguan rebels by secretly selling arms to Iran, in contravention of U.S. policy, and diverting the proceeds to the Contras. Both President Reagan and Vice President Bush explicitly denied that they knew about the Iran-Contra operation. Admiral John Poindexter, the President's National Security Adviser, testified to a congressional investigating committee that Reagan's ignorance was the result of a deliberate decision by Poindexter.

The independent counsel investigating the Iran-Contra affair concluded that "President Reagan was apparently unconcerned as to how details of his policy objectives for Contra support were being carried out by subordinates who were operating virtually free from oversight or accountability."

With respect to Vice President Bush, who had had more experience in these matters than anyone else at the NSC, the independent counsel concluded: "Bush acknowledged that he was regularly informed of events connected with the Iran arms sales.... These statements conflicted with his more extreme public assertions that he was 'out of the loop' regarding the operational details of the Iran initiative and was generally unaware of the strong opposition to the arms sales by Secretary of Defense Weinberger and Secretary of State George P. Shultz. He denied knowledge of the diversion of proceeds from the arms sales to assist the Contras. He also denied knowledge of the secret contra-resupply operation...."

Congress

In the beginning, such oversight of the intelligence community as Congress exercised was carried out by the Armed Services Committee and Appropriations Committee of each house. This jurisdiction devolved on Armed Services because the CIA had been created by the National Security Act. A further consideration was that most intelligence work is done by agencies in the Defense Department. Jurisdiction devolved on the Appropriations Committees because they control the money.

None of these committees had much stomach for the task. They operated through subcommittees that had little staff support, if any, and rarely met. Senator Leverett Saltonstall, for years the ranking minority member of the subcommittee in the Senate, once remarked that there was information "which I ... would rather not have, unless it was essential for me as a Member of Congress to have it...."

This was not an attitude that inspired sharp questioning. It was, however, the attitude that prevailed for thirty years. During this time, congressional opinion gradually shifted to the view that Congress really ought to pay more attention to what the intelligence agencies were doing. The main factor in this shift was a series of spectacular intelligence failures or misdeeds. Contributing factors were growing public disaffection with government stemming from the Vietnam war and the Watergate scandal, and revelations in the news of abuses of the rights of American citizens. Among the events that influenced Congress were:

- The U-2 affair in 1960, which is mentioned briefly above.
- The abortive Bay of Pigs invasion in 1961. Ex post facto reviews, both in the executive branch and Congress, revealed that the CIA-organized expedition had proceeded on the basis of seriously flawed intelligence, which overestimated anti-Castro sentiment in Cuba. The Bay of Pigs also breached the doctrine of plausible deniability; President Kennedy took full responsibility.
- The invasion of the Dominican Republic by President Johnson in 1965, ostensibly to restore order in the aftermath of an attempted coup and to forestall a communist takeover. This was not so much a failure of intelligence as an abuse of intelligence by a President who exaggerated intelligence reports, both of public disorder and the communist threat, in order to justify his actions.
- Press revelations in 1967 of secret CIA subsidies to the National Student Association and the Institute of Political Education. In the former case, the subsidies paid for delegations to international meetings so that American voices could be heard. The Institute of Political Education, which was located in Costa Rica, was sponsored by liberal democratic parties in several Latin American countries and was paid for by the CIA with the purpose of electing Juan Bosch President of the Dominican Republic after the assassination of the longtime dictator Rafael Trujillo. Most Americans would probably have supported both of these efforts; as a matter of fact, work similar to that of the institute is now carried on quite openly in a number of countries by publicly funded adjuncts of the Republican and Democratic Parties. But the context of the revelations left the impression that the CIA was meddling for some nefarious purpose.
- Efforts of American business and the CIA to destabilize Chile during the presidency of the Socialist Salvador Allende (1970–73). In the 1970 election, the prospect that the Chilean congress might choose Allende, the front-runner, so alarmed the International Telephone and Telegraph Company, a heavy investor in Chile, that it offered the CIA $1 million toward the cost of a covert action to subvert the congressional process. Richard Helms, the CIA director, declined the offer, but it became public through a leak to the press and a public uproar ensued. The issue was rekindled in 1973 when Allende was deposed and died in a military coup. The role of the CIA in this, if any, was and remains

murky, though there was no doubt about the Nixon administration's dislike of Allende. Congress subsequently explicitly authorized the CIA to accept gifts. This opened the way to CIA collaboration with private entities—a singularly ill-advised policy.

- Finally, in December 1974, Seymour Hersh of the *New York Times* published a lengthy report of domestic activities by the CIA. These included intelligence files that were maintained on thousands of American citizens, surreptitious mail openings, break-ins, and wiretapping. Most of these violated a provision of the National Security Act expressly forbidding any internal security functions by the CIA. President Ford appointed a committee under Vice President Rockefeller to investigate, and both the Senate and the House had their own investigating committees. It was out of these investigations, under Senator Frank Church and Representative Otis Pike, that the permanent intelligence-oversight committees came to be.

INTELLIGENCE COMMITTEES

The intelligence committees are structured somewhat differently from other congressional committees. As Robert Dreyfuss pointed out in Chapter 8, some of the differences are technical but have important operational consequences. The committees are "select," not standing. This means that the members are appointed by the leadership, not by party organizations. It is a way of strengthening leadership control.

Term limits are also in effect: eight years in the Senate; four terms (eight years) in any six consecutive congresses in the House. The rationale for this innovation was to guard against co-option of oversight committees by the agencies that they are supposed to be overseeing. It was thought that periodically changing the membership would mean that the agencies would have to start over in their campaigns to co-opt their overseers. It could just as well be argued, however, as some former members of the Intelligence Committees forcefully do, that term limits cut short a member's useful service after he or she goes through an apprenticeship of familiarization.

There is a countervailing consideration. As term limits revolve members through the Intelligence Committees, they increase the general level of congressional knowledge of the intelligence community. This is a good thing.

Term limits do not apply to staffs. Some staffs in both houses have served longer than members, a practice that tends to give the staff more influence than members (or more influence over members), a situation of tails wagging dogs.

Some members of each Intelligence Committee are drawn from other committees with intelligence interests—Appropriations, Armed Services (called National Security in the House), Foreign Relations (International Relations in the House), and Judiciary.

Finally, there is a matter of jurisdiction that should be noted. In the Senate, the Intelligence Committee has jurisdiction over the NFIP (the National Foreign Intelligence Program), and the Armed Services Committee over TIARA (Tactical Intelligence and Related Activities). In the House, jurisdiction over the NFIP is in the Intelligence Committee, but jurisdiction over TIARA is shared by Intelligence and National Security. This not only makes congressional procedures more cumbersome, but it embeds in Congress some of the same jurisdictional rivalries found in the executive branch. The result is that on some issues, congressional committees tend to become advocates of particular agencies or programs instead of overseers.

Power of the Committees

The basic, ultimate power of the Intelligence Committees comes from their control of funds for the intelligence community. Each year Congress passes an Intelligence Authorization Act, which authorizes stated amounts of money to be appropriated for specified intelligence agencies and programs. These Authorization Acts are written in the Intelligence Committees, with the President's budget requests as the starting point. They are followed by bills from the Appropriations Committees. The appropriations may be less than the authorizations, but not more.

The Authorization Act is the most important thing the Intelligence Committees do each year, and the process takes several months. Almost everything about it is secret: the budget requests, the committees' actions, the law that eventually emerges from the legislative process. The actual amounts of money are specified in secret appendices of the bills and committee reports. These appendices are sent to the Office of Management and Budget and to the agencies concerned.

Committee reports shed some light on how the committees view the intelligence community and its problems, at least in general terms. The reports reflect the committees' judgment as to where the community and its executive overseers in the White House ought to put more attention. But the total budget is so huge—lately in the neighborhood of $27 billion to $30 billion a year—that a congressional committee cannot consider it in detail. This applies also to other segments of government spending, such as the Defense Department and other large agencies. Congressional oversight of the intelligence budget is probably not much better or worse than congressional oversight of the budget for the Defense Department, which is, of course, even larger.

A more important question is why so much secrecy attends the process. It is understandable and defensible that detailed intelligence expenditures should be secret. One would not want the government of country X to know how much money the United States was spending in an effort to suborn its leaders. But surely no foreign adversary could learn anything useful from a statement that the grand total authorized by an intelligence bill was $26.6 billion. That was the amount spent in fiscal year 1997, announced on October 15, 1997, by Director of Central Intelligence George Tenet in response to a lawsuit filed by the Federation of American Scientists.

It is a common practice for a committee report to say that a particular year's authorization is X percent more or less than the preceding year's. When Tenet announced that 1997 spending was $26.6 billion, Representative Norman D. Dicks said that the authorization bill for fiscal year 1998 provided 1.4 percent more. It follows that the 1998 budget was $27 billion.

Some of Tenet's predecessors, going back to Stansfield Turner in the Carter administration, said that they had no objection to publication of an overall figure for intelligence spending. Yet Congress has repeatedly voted against this. Sundry arguments are made in favor of keeping the total secret: It would be the first step down the slippery slope toward breaking down the overall figure; publication of totals over a number of years would disclose trends in intelligence spending that would provide clues to foreign adversaries; it would be argued that Congress could not intelligently debate total figures without knowing the components. (The fact is that Congress does know the components; any member can go to the offices of the Intelligence Committees and review the budget in exhaustive detail. Not many members do.)

Rarely stated is the consideration that if the public knew the extent of intelligence spending, demands to curtail it would increase. Another argument, not much used, is that as the Defense Department's Cold War budget shrinks, it is becoming more difficult to hide the intelligence component.

The power of the purse is the most basic of all congressional powers. Congress can effectively block any activity of which it disapproves, intelligence or otherwise, by a simple provision stating, "Notwithstanding any provision of law, no funds may be expended for X." That is how Congress finally ended U.S. participation in the Vietnam war. That is how Congress ended, it *thought,* aid to the Nicaraguan Contras.

Given the fact that, so far as is publicly known, the intelligence budget has been relatively stable in recent years, Congress may not appear to have been very diligent in exercising the power of the purse. But that does not necessarily follow. Some intelligence operations that are large in controversy are small in expenditures, so a particular operation could be blocked without having a noticeable impact on the budget.

Most such operations are covert actions, which are subject to special procedures. The law requires that covert actions be reported to the Intelligence Committees before they are undertaken, and that these reports be accompanied by presidential findings that the actions are important to the national interest. This requirement marked the effective end of plausible deniability. The heart of the Iran-Contra scandal was the effort of the Reagan White House to evade the presidential finding and the report to Congress.

When the President has approved a covert action, the DCI appears before the committees to explain and justify it. Any member of the committee can raise a question or objection. Sometimes these are so serious that the DCI goes back to the President with a recommendation that the action be reconsidered. On some occasions, the chairman and the ranking minority member of the committee call on the Secretary of State, the National Security Adviser, or even the President himself to express strong objections. According to persons who have been involved in this process, it has sometimes led to modification or even cancellation of a proposed covert action.

If the President decides to go ahead with a covert operation despite objections, Congress has two other ways to stop it: Withhold money in the next appropriation cycle, or pass a bill expressly forbidding the action. We do not know how many times, if any, Congress has taken the first route. Congress has taken the second twice, once with respect to Angola and once with respect to Nicaragua. It was the Reagan administration's violation of the Nicaragua ban that led to the Iran-Contra scandal.

Access to Information

In order to exercise oversight of covert action or anything else, Congress has to know about it—everything about it. Despite the plain language of the law requiring information to be made available, the intelligence community has steadily resisted full reporting. This tends to follow a pattern. The community volunteers nothing. When asked, it tells a little. When pressed, it tells a little more. When pressed further, it tells a little more yet. In extreme cases, the President will assert what he calls his constitutional powers as grounds for nondisclosure. If he is adamant, no matter how specious the claim, Congress has the choice of backing down or cutting off the money. Matters rarely reach this stage. If an Intelligence Committee persists, it usually gets what it wants, though sometimes the information is limited to the chairman and ranking minority member. How much a committee gets is a function of its persistence, and that, in turn, is a function of its political will.

The obstinacy of the intelligence community in withholding information from Congress is one of the principal roadblocks to effective oversight, as well as to better Congress–executive-branch relations. Executive agencies never seem to learn that they are better off supplying information graciously in a timely fashion than grudgingly after long delay. But on the same point, Congress has been too ready to accept qualifying phrases, such as "to the extent consistent with the constitution," in legal requirements to give it information.

The intelligence community's objections to sharing information are rooted in tradition, in the culture of the community (especially the covert-action division of the CIA), and in some practical reasons. The practical reasons are both legal and operational.

Legally, the National Security Act places upon the Director of Central Intelligence the responsibility of protecting intelligence sources and methods. The necessity for doing this is self-evident. If sources are known, they can be silenced—killed or otherwise muted if human, or blocked in one way or another if technical. If methods of collection are known, steps can be taken to render them ineffective.

Sources or methods are sometimes disclosed, perhaps inadvertently, through disclosure of the intelligence that has been gained from them. A particular piece of intelligence might be known to only a few people in the government of country X. Disclosure that the United States knows this intelligence will enable country X to narrow the field of possible sources and perhaps identify the culprit. Or, if the intelligence could have come only from an intercepted radio or telephone communication, disclosure will tell country X that the United States is intercepting its messages. If the message was sent in code, it means that the United States has broken the code. Country X will then change its codes, and the United States will have to begin again the laborious task of code-breaking. This is precisely what happened when the Reagan White House itself released decoded Libyan messages establishing Libyan complicity in bombing a Berlin nightclub frequented by American soldiers.

The operational reason for the intelligence community's reluctance to share information is that most intelligence activities, if publicly known, become counterproductive. For instance, had the CIA revealed its subsidy of noncommunist labor unions in Western Europe after World War II, the unions could have been dangerously compromised. In various countries, the CIA has at one time or another paid newspapers to print articles that the Agency had prepared. Public knowledge of this would have robbed the newspaper of whatever influence it had. One might argue that the CIA should not be engaging in these activities in the first place, and that revealing them and making them ineffective therefore does no harm. That misses the point. If it is thought that the CIA should not be doing something, the President or Congress should have blocked it before it started. If something is going to be done, it ought to be done in a way that offers hope of success. Another consideration: If a collaborator is revealed after having been promised confidentiality, it becomes harder to recruit other collaborators.

The most important reason that the intelligence community resists sharing information with Congress is distrust. The executive branch in general, and especially the intelligence bureaucracy, does not trust Congress to keep secrets. This is a little like the pot calling the kettle black. Close observation of unattributed news stories containing intelligence information suggests that more leaks come from the executive branch than from Congress.

Distrust runs both ways. Congress does not trust the executive branch not to exaggerate intelligence, or even not to tamper with it for political purposes. Thus, when the White House cites intelligence that cannot be revealed as justification for a controversial policy, Congress immediately thinks that the facts do not support the policy. After the bombing of a Sudanese pharmaceutical factory in August 1998, it was congressional and news-media skepticism that drove the Clinton administration to reveal, reluctantly, that secretly taken soil samples from the plant contained elements of nerve gas. Even this did not entirely end the controversy.

Nonetheless, executive-branch uneasiness about sharing secrets with Congress is understandable. Under the rules of both houses, information available to any committee is available to any member; all the member has to do is show up at a committee's offices and ask for it. The intelligence community argues, with some reason, that although most members of Congress are trustworthy, a few may not be. Some of the CIA's activities in Chile in the early '70s were revealed by Representative Michael Harrington, who was not a member of the Armed Services Committee but had read a transcript of testimony by Director of Central Intelligence William Colby.

The case of Representative, now Senator, Robert G. Torricelli of New Jersey is illustrative of how far the intelligence community, with support from the Justice Department and the White House, will go to make it more difficult for Congress to obtain information. After the murders of Michael Devine, a U.S. citizen, and Efrain Bamaca Velasquez, a Guatemalan rebel leader married to an American citizen (see Chapter 2), responses to congressional inquiries by the American embassy and the CIA station in Guatemala proved to be misleading. In the embassy's case, this was because the embassy itself had been misled by the CIA. In 1995, Richard A. Nuccio, a State Department employee, told Torricelli, a member of the House Intelligence Committee, that there was evidence linking Colonel Julio Roberto Alpirez, a CIA Guatemalan informant, to the killings. Torricelli gave the story to the *New York Times*.

This got Torricelli in trouble with both John Deutch, the DCI, and Speaker of the House Newt Gingrich. Gingrich demanded that Torricelli lose his seat on the Intelligence Committee, but Richard A. Gephardt, the House minority leader, who controlled Democratic appointments to the committee, refused to act. The case was referred to the House Ethics Committee, which concluded that the rules were sufficiently ambiguous to preclude action against Torricelli.

Separate investigations by the CIA Inspector General and the President's Intelligence Oversight Board came to conclusions that, on their face, were contradictory: No CIA employee had "knowingly misled the congressional oversight committees or deliberately decided to withhold information from them," but, at the same time, the agency had not met its responsibilities to keep the committees informed. Deutch forced two senior CIA officials in the Directorate of Operations—the former chief of the Latin American Division and the Station Chief in Guatemala—into early retirement. He downgraded or sent letters of reprimand to seven others. Deutch also revoked Nuccio's security clearance, despite the fact that Nuccio worked for the State Department and not the CIA.

Torricelli was elected to the Senate in 1996 and became a member of the Intelligence Committee. In 1997, that committee put a provision in the intelligence authorization bill for fiscal year 1998 requiring the President to make it clear to federal employees that they can inform Congress of wrongdoing without fear of reprisal, even when the evidence involves classified information. In its report on the bill, the Intelligence Committee said it was concerned that federal employees would get a signal that "there are no circumstances" in which they can disclose evidence of wrongdoing involving classified information. This provision passed the Senate but was dropped in conference in the face of strong House opposition.

The Senate committee tried again in 1998. Over strong objections from the administration, it unanimously approved a bill requiring the President to inform employees of the CIA, the DIA, NIMA, the NSA, and the FBI that they can disclose to members of Congress on relevant oversight committees any violation of law, false statement to Congress, or other abuse. The Justice Department argued, preposterously, that the bill would violate the separation of powers. This result would follow, the Justice representative said, by enabling any employee with access to classified information to determine the circumstances for sharing that material with Congress, regardless of the effect on the President's decision-making on national security and foreign affairs. In the fall of 1998, a compromise bill was passed by both houses, and the administration did not oppose it. It went into effect as part of the Intelligence Authorization Act for fiscal year 1999.

The administration's argument made it clear that it wishes to determine unilaterally the circumstances for sending information to Congress. This was unacceptable. The right to control the flow is the right to control the content. The position reflected a traditional reluctance to accept Congress as a participant in the policymaking process. The congressional role, in this view, should be limited to passing upon, possibly with changes, policies and programs earlier formulated in the executive branch and put forward with the presidential imprimatur. Thus the executive branch does not like to tell Congress anything about an emerging problem until its own (that is, the President's) position has cleared a tedious interagency coordination process with every "i" dotted and every "t" crossed. In bureaucratic jargon, nothing is fit for Congress to know "until we have our ducks in a row and everybody is on board."

This position has made Congress feel less like even a junior partner than like an inquisitive child waiting to have something explained to him—or even waiting until he's old enough to know. It has caused more resentment in Congress than the bureaucracy is probably aware of. Yet not even Presidents who have come from Congress (Truman, Kennedy, Johnson, Nixon, Ford, and Bush) have shown much understanding of it.

In 1996, it was disclosed that, through failing to object, the United States had given tacit approval to Iranian arms sales to Bosnia. Congress smelled an unreported covert action, something that would be guaranteed to raise hackles on Capitol Hill. What had happened, however, was not an action but a failure to act. The Intelligence Oversight Board investigated and concluded that the U.S. role had not been a covert action requiring a presidential finding. The White House denied a request by Senator Arlen Specter of Pennsylvania, then chairman of the Senate Intelligence Committee, for access to the report, on grounds of protecting confidential policy discussions. Republicans were infuriated; Democrats probably did not like it, but kept quiet.

Although the White House goes about defending its perceived powers in an impolitic way, there is understandable reason for its concern. The executive branch is chaotic enough as things stand. Mid- and upper-level officials actively seek contacts on Capitol Hill, as well as in the news media, that they think might be helpful in interagency squabbling. Congressional committees investigating an executive agency get some of their most valuable tips from employees of that agency. The most infamous example of this occurred in the anticommunist crusades and demagoguery of Senator Joseph R. McCarthy in the early 1950s. On a different level, an anonymous letter from a Navy officer provided the first indication to the Senate Foreign Relations Committee that the White House and the Defense Department had not told the whole truth about the Tonkin Gulf incidents in the summer of 1964.

The White House would naturally prefer that the government over which it presides be more orderly, but from the point of view of the public interest, order and efficiency are not necessarily what are needed. The government was not designed to be orderly and efficient. Members of Congress and their staffs can reasonably be expected to sort out the useful from the meretricious and mendacious in the information they receive from executive-branch visitors. Some irresponsible members do not do this, and that then becomes a matter for Congress to handle—as the Senate did, after dallying too long, in the case of McCarthy.

The proposal that some members of Congress can have certain information and others cannot goes against the grain. It amounts to making some members more equal than others. That implies that the people of some states and congressional districts are entitled to more representation than others. This is contrary to the traditions of Congress and to the Constitution.

It should be pointed out that access to information is by no means a problem unique to the Intelligence Committees. In varying degrees, it is faced by every committee in Congress. A government agency rarely says no to a congressional request. It simply doesn't say yes to a request that it does not wish to answer. The Departments of State and Defense are past masters at delaying indefinitely without admitting that they do not intend to comply.

It should also be pointed out that the public interest was served by the disclosures of Harrington and Torricelli. Harrington's story led to further disclosures of the Nixon Track II campaign against Allende in Chile; Torricelli's led to the unraveling of previously unknown involvements in Guatemala. It might have been better if Torricelli had first taken his concerns to the House Intelligence Committee, of which he was a member—but Harrington did first take his concerns to both the Senate Foreign Relations Committee and the House Foreign Affairs Committee, without results.

Micromanagement

It is hard to draw the line between issues of public policy and matters of administrative detail. Whether a particular covert action should be undertaken is a matter of public policy; the operational details involve micromanagement. Whether to shift intelligence spending from human sources (usually comparatively cheap but not always reliable) to technical sources (very expensive but foolproof, if they work) is a policy question. Which human sources to recruit and how to design reconnaissance satellites are micromanagement.

Even the question of what kind of human sources to use is a blurred area between policy and micromanagement. Congress has placed the Peace Corps and Fulbright scholars unequivocally off-limits to the CIA; there is controversy about other groups of Americans abroad—non-

Fulbright academics, clergy, and especially journalists. In 1996, the House, by a vote of 417 to 6, approved an amendment to the 1996 intelligence authorization bill prohibiting the use of U.S. journalists in espionage operations, but the conference committee added a presidential waiver. This is the same as saying, "Don't do it unless you really want to." Congress is fond of imposing sweeping restrictions on the executive branch and then adding something to the effect that "unless the President finds that whatever has just been prohibited is vital to the national interest or national security." Since it is characteristic of Presidents to think that anything they want to do is vital to the national security, this vitiates the restriction.

There are policy reasons for the CIA not to recruit academics, clergymen, or journalists. If the intelligence cover of one member of a group is revealed, all members of that group are tainted with suspicion of espionage. On the other hand, the more Congress narrows the pool of potential agents, the more it limits the pool of CIA sources and the cover available for CIA officers. It is in the interests particularly of journalists to maintain a certain distance from the CIA; among other reasons, the ethics of the profession require it. A journalist can say no to an attempt at recruitment without Congress passing a law. But there are a variety of mutually beneficial relationships that stop short of employer and employee.

There has been concern about CIA recruitment of foreign sources who come from distasteful backgrounds or occupations. This was part of the criticism of the relationship the CIA had with Colonel Alpirez in Guatemala. Here is a dilemma. Both Congress and the White House expect the CIA to know what the military is up to in a country like Guatemala, and the CIA can't very well know this without military informants. However, in some countries, such as Guatemala, the relationship with an informant may include financial or other active forms of support, constituting a relationship that may seriously compromise the moral integrity of the Agency. Even if a relationship falls short of active support, the informant may well think that his connection with the CIA amounts to a U.S. stamp of approval, giving him added prestige among his peers.

Co-Option

President Eisenhower warned against the influence of the military-industrial complex: the alliance between the armed forces and the corporations that manufacture their weapons. As Robert Dreyfuss points out, these two groups are dependent on each other and thus have a common interest in influencing the Armed Services Committees of Congress. With the intelligence community increasingly dependent on reconnaissance satellites and other expensive technical means of collection, there has grown up an intelligence-industrial complex with a vital interest in the annual intelligence authorization bills, and therefore in influencing the Intelligence Committees.

The Intelligence Committees have been particularly susceptible to revolving-door staffs—that is, staffs that come to the committee from the intelligence community or vice versa. When John M. Deutch became Director of Central Intelligence in 1995, he appointed as his deputy George J. Tenet, who had been majority staff director of the Senate Intelligence Committee from 1988 to 1992. Deutch, who had previously been Deputy Secretary of Defense, brought with him a number of other persons with congressional experience:

- Nora Slatkin, formerly on the staff of the House Armed Services Committee, to be Executive Director of the CIA;
- Keith R. Hall, formerly on the staff of the Senate Intelligence Committee, to be director of the intelligence community management staff;
- Jeffrey H. Smith, former counsel of the Senate Armed Services Committee, to be general counsel of the CIA; and
- Michael J. O'Neil, former chief counsel of the House Intelligence Committee, to be Chief of Staff.

After Tenet succeeded Deutch as DCI, he appointed Britt Snider, who had served with him on the Senate Intelligence Committee staff, to be the CIA's Inspector General.

There are also examples of movement the other way. Representative Porter Goss, who had been a clandestine-services officer of the CIA before being elected to Congress, became chairman of the House Intelligence Committee in 1997. On the staff level, Barry Goldwater, chairman of the Senate Armed Services Committee, brought John Blake, who had capped a thirty-two–year CIA career as acting Deputy Director, to be staff director of the committee. Blake was succeeded in that position by Robert R. Simmons, a ten-year CIA veteran.

When the staff—or members, for that matter—move in either direction between Congress and an executive agency, they naturally bring with them institutional biases, but they also bring a familiarity with the agency or committee they are coming from.

Revolving-door staffs are by no means a phenomenon limited to intelligence. The Senate Armed Services Committee at one time had a retired major general as staff director. The Senate Foreign Relations Committee has a long record of recruiting staff from the executive branch and of staff members moving to the executive branch.

The members and especially the chairmen of committees have a great deal to do with whether co-option occurs. Some chairmen are predisposed to defer to the executive branch; others are not. Before he became chairman of the Senate Intelligence Committee, Senator Goldwater expressed the view that the committee ought not even to exist: "The Russians have a very fine system.... No part of their government has any idea of what is going on [in the KGB]. That is the way ... I wish it were in our country, but it is not."

Despite his aversion to oversight, Goldwater was publicly outraged over the CIA's mining of Nicaraguan harbors in 1984 without telling the Intelligence Committee. This, at least in the view of Senator Daniel Patrick Moynihan, led the CIA to mount an anti-Goldwater whispering campaign, spreading gossip about how the Senator was getting old and his memory was fading.

Problems of Congressional Oversight

In general, Congress's mechanisms and procedures for effective oversight are already in place. It could tighten legal provisions about its right to know and about timeliness in reports to Congress. It could usefully pay more attention to the intrusive methods of counterintelligence. Much of this is carried out in the United States and therefore potentially involves infringement of the rights of American citizens. This makes it at least as much a domestic as a foreign issue and involves the Judiciary as well as Intelligence Committees.

There have been various suggestions about enacting legal guidelines or even a basic legislative charter for the intelligence community. Such a charter was one of the recommendations of the Church Committee, and the Intelligence Committees have seriously considered it from time to time, notably in 1980, when the Senate committee held extensive hearings and actually produced a bill—which was not passed.

Piecemeal, however, there has been some accomplishment. The Foreign Intelligence Surveillance Act takes a step toward extending the safeguards of the Fourth Amendment to protect citizens from surreptitious wiretapping and other intrusive searches and seizures. The act created a special court of seven U.S. district judges, appointed by the Chief Justice, to hear applications for electronic surveillance. An appellate procedure is provided, but since the court meets in secret and its orders are secret, an aggrieved party does not know any of this. The congressional committees could review how well this act has been working and possibly take further steps toward reinforcing individual rights.

The Classified Information Procedures Act dealt with the problem of "graymail," a technique by which a defendant had been able to frustrate prosecution by demanding the production of classified documents that he claimed were essential to his defense. The act provides that the government may move to suppress certain information, and that the court is to rule on this motion in a secret, ex parte hearing. The act had an unintended effect in prosecutions brought by independent counsels who are not responsible to the Justice Department. If the Justice Department is hostile to the prosecution, as was the case in the Iran-Contra cases, then the department becomes in effect a part of the defense. In this circumstance, an act that was supposed to strengthen the prosecution strengthens the defense instead. This could well be reviewed.

The Intelligence Identities Protection Act was designed to prevent the identities of CIA officers stationed abroad from being made public. It was passed in response to a series of attacks (one of them fatal) on such officers in various countries. It avoided running afoul of the First Amendment by directing the prohibition to people who learned the information during the course of their official duties or who engaged in "a pattern of activities ... with reason to believe that such activities would impair or impede the foreign intelligence activities of the United States." This seems an unobjectionable precaution.

The annual Intelligence Authorization Acts have been the vehicle for steps toward reorganizing the intelligence community, mainly by strengthening the authority of the DCI. The Authorization Act for fiscal 1993 created the National Intelligence Council in the office of the DCI and made it the community's principal organ for analysis. Congress has made a stab at increasing the authority of the DCI over the community's budget, over reprogramming funds, and over relations with foreign intelligence agencies. The DCI has also been given a say in the appointment of the heads of intelligence agencies in the Department of Defense, although congressional half-measures did not change the essential dynamic described by Richard Stubbing—that the DCI still has no authority over budgets or personnel beyond the CIA, despite the creation in 1997 of a Deputy Director of Central Intelligence for Community Management. The appointments are still made by the Secretary of Defense, but in consultation with, or on the recommendation of, the DCI.

As things now stand (and have stood since almost the beginning), approximately 80 percent or more of the intelligence budget goes to agencies in the Defense Department. Successive DCIs trying to assert either budgetary or operational control have encountered, and no doubt will continue to encounter, stout resistance from the Secretary of Defense.

Before Congress can get serious about reorganizing the defense and intelligence communities, it has to do something about reorganizing itself. A principal problem is jurisdictional rivalry between the Intelligence and Armed Services Committees, analogous to the bureaucratic jealousy between the CIA and the Defense Department. The Judiciary Committees, which have jurisdiction over the FBI, also have to be taken into account with respect to counterintelligence. The matter is further complicated by the tendency of oversight committees to view themselves as protectors of the agencies they are overseeing.

Such coordination as exists is exercised in the White House by the Office of Management and Budget. This is the President's tool for managing the government. If the President wants a truly centralized intelligence service—under the Director of *Central* Intelligence, as Truman intended—he can get it through the OMB, but at the cost of alienating the Defense Department and the military services.

Congress could also order centralization through statute, but Congress is paralyzed in this respect by the same divisions that paralyze the intelligence community. There is a question of the degree to which analysis ought to be centralized. Intelligence data are frequently incomplete, ambiguous, and susceptible to conflicting interpretations. Separate, independent analytical processes can better serve policymakers even when they produce conflicting reports.

Meanwhile, there are other pieces of legislative oversight that Congress might usefully address. Covert action is prominent among these. Congress has already enacted elaborate requirements for reports to Congress relevant to covert actions. Should it go further and legislate what is and is not permissible with respect to tactics and objectives?

Here seemingly insuperable difficulties arise. Different members of Congress—and of the public, for that matter—react differently to the infinite variety of international situations in which the United States can find itself. Some people feel that some covert actions are good; others think they are bad. Objectives and tactics are important, but not necessarily crucial, determinants.

Assassination is ruled out by long-standing executive order. But after the terrorist attacks on American embassies in Kenya and Tanzania in August 1998, L. Paul Bremer III, who had been President Reagan's ambassador-at-large for counterterrorism, called for lifting the ban.

If there is to be any other restriction on covert action, banning the use of military or paramilitary forces could well be considered. The introduction of such forces enlarges an activity so much that it can rarely be kept secret, and publicity generally destroys a covert action. That rule did not hold, however, in the case of Afghanistan, when covert assistance to the anti-Soviet resistance in the 1980s was well known.

The mischief of covert action, as stated at the opening of this chapter, is the temptation it presents to the President to accomplish in secret something that would embroil him in political controversy if attempted openly.

That is the aspect that most needs congressional oversight, but it is not susceptible to a legislative solution. One helpful addition to the law, as has been suggested above, might be a requirement that the President include a risk-benefit analysis in his finding justifying a covert action. It might be a good idea also to levy such a requirement on the interagency committees that consider major intelligence operations at the under secretary or assistant secretary level.

Finally, one of the intelligence community's problems, even in the Cold War, was selecting targets and priorities. These problems, as noted above, are even greater now, and Congress ought to provide direction.

CONCLUSION AND RECOMMENDATIONS

Almost thirty years ago, a reluctant Congress was driven to more detailed oversight by a series of intelligence failures and by the sense that the intelligence community, especially the CIA, was out of control. Senator Church once called the CIA a "rogue elephant." He later learned better, but that was an image that influenced the framework of early oversight efforts. In fact, most of the CIA's sins have been committed by stretching orders from higher authority and by an excessive obsession with secrecy; an example is the failure to tell the American Ambassador to Guatemala about the Agency's relationship with Guatemalan officers involved in human-rights abuses. The emphasis of the oversight committees has consequently been to prevent failures and scandals. Its record has not been encouraging.

Faulty analysis allowed the government to be surprised by the overthrow of the Shah of Iran. The moral of this story is that public sources are frequently better than secret ones (and a good deal cheaper). The newspapers that are read daily in the White House contained plenty of indications over a period of months that the Shah's fall was imminent.

The Intelligence Committees allowed themselves to be deceived with respect to the Iran-Contra scandal, which came to public attention only through a series of coincidences. The committees did ask what was going on with respect to the Contras, but they did not ask enough people.

The intelligence community failed to predict the end of the Cold War and the collapse of the Soviet Union. The principal argument within the community and within Congress before these cataclysmic events concerned whether the prevailing wisdom *underestimated* Soviet strength.

The difficulty of legislating in this field is the strongest argument for critical oversight. It cannot be expected that oversight, no matter how meticulous, will prevent all mistakes, or outcomes that some people, perhaps many, will think deplorable. But there are likely to be fewer mistakes with oversight than without it. This is why the Constitution provides for the separation of powers. There is no guarantee that the President and Congress will not both be mistaken at the same time; the premise of the Constitution is only that they are less likely to be.

However, if the system is going to work, both the executive and legislative branches need to be fully involved. For a variety of reasons, the legislative branch has not been. The most important of these reasons is lack of political will. The authority Congress needs is in place. Using it requires determination and persistence. A staff member of one of the Intelligence Committees once explained it: "There is a marked lack of curiosity around here." After more than forty years, there are still too many members who, like Senator Saltonstall, would just as soon not know.

Given these considerations, the following recommendations seem in order:

1. Policymakers, both in the executive branch and Congress, should subject proposed intelligence operations to more rigorous cost-benefit analyses. Will the intelligence or other policy goal expected from an operation be worth the expected cost, including the cost of possible failure?
2. Plausible deniability should be ruled out as a condition of any intelligence plan.
3. Congress should repeal the authority of the CIA, or any other intelligence organization, to accept gifts. Gifts should be positively prohibited. If an intelligence operation is worthwhile, the government should pay for it, not be beholden to outside donors.
4. Certain rules should be changed to make the Intelligence Committees like other committees:

    a. They should be standing, not select, committees.
    b. Party leaders should not have ex officio status as members of the Intelligence Committee in the Senate.
    c. Service on Intelligence Committees should not be subject to term limits.

    The effect of these changes would be to make the committees more representative of the House and Senate as a whole and less reflective of the leadership.
5. The Intelligence Committees should be given jurisdiction over all components of the intelligence community, including not only the CIA but the TECHINT agencies as well. This would mean reducing the jurisdiction of several other committees, mainly Armed Services and Judiciary. It would be politically difficult, but it would enhance the authority of the Intelligence Committees and facilitate a needed reorganization of the intelligence community.

6. More intelligence budget data should be made public. The information should unquestionably include the overall figure, and probably the total for each intelligence agency as well. The burden should be on each agency to show why it shouldn't publish its budget data.
7. Congress needs to review all intelligence legislation with special reference to:

    a. eliminating loopholes in reporting requirements;
    b. tightening protection of civil liberties in surveillance authority;
    c. reviewing the Classified Information Procedures Act to avoid ex parte procedures.

8. Consideration should be given to banning paramilitary covert action.

CRAIG EISENDRATH

Conclusions

THE CENTRAL QUESTION raised in this book is this: In the post–Cold War world, what kind of intelligence system is essential for our security and appropriate to our democratic society? The book has been motivated by a desire to achieve the best possible intelligence system, one that serves the national interest and does the least possible harm here and abroad.

As these pages make clear, the system we currently have does neither. It fails to provide the kind of information U.S. policymakers need to promote the national security, and it does unnecessary damage to people both in this country and other nations, and to our international reputation. Instead of thinking of the United States as the country of Jefferson and Lincoln, too often foreign people equate us with the machinations of the CIA. Here at home, the intelligence system erodes freedoms that are at the heart of democracy, and legitimates political behavior that corrupts it.

Each of our authors has ended his or her contribution with specific recommendations. The purpose of this chapter is to draw these recommendations together and present a coherent outline for reform. Although the authors have their own points of view, their remarkable agreement on the main lines of reform makes such an outline an organic development from the text.

As these recommendations make clear, a major overhaul will be needed if our intelligence system is to do its proper job and if we are to achieve a reputation among the nations of the world commensurate with our democratic ideals.

INTELLIGENCE IN THE POST–COLD WAR PERIOD

The first requirement for reform is to be clearheaded about what need intelligence should fill. The paramount need, according to Roger Hilsman and others in this study, is to supply the President and Congress with information about possible strategic threats to the United States. If we survey the world region by region, as Hilsman does, we find specific places from which such intelligence is required; such a survey suggests an important mission for intelligence, but a limited number of targets.

Hilsman finds that the world still harbors governments, such as those of Russia, China, North Korea, Iraq, and Iran, that, if not capable of threatening the existence of the United States, are capable of inflicting damage. Terrorist organizations pose dangers as well. Intelligence is thus needed on possibly threatening states, terrorism, and the proliferation of nuclear, biological, and chemical weapons. There is also a continuing requirement within the U.S. government for centralizing intelligence, unifying the work of seasoned analysts and outside experts, and using all sources, overt and covert, to provide an accurate picture of the world for policymakers.

As the authors of this book make clear, there exists today no power that directly threatens our existence, as did the Soviet Union, and consequently there is no present moral or practical justification for the extremity of illegal and nondemocratic actions that the United States practiced during the Cold War. But beyond this, case after case presented by the authors states eloquently that these actions, even during the Cold War, were mainly ineffective and often counterproductive.

The authors maintain that information secured by clandestine means was frequently unreliable, and in particular that paid sources of intelligence were usually more concerned with pleasing their "handlers" than with telling the truth. Covert actions cost the United States dearly in reputation, frequently ended in failure, caused unnecessary suffering in the countries in which they occurred, and had harmful effects within the United States as well.

This book emphasizes the need to deal with most of the world's problems—perhaps almost all of them—not through the risky expedients of espionage and covert action but through the solid methods of statecraft. The authors argue that the option of espionage should be exercised only when reliable, open information cannot be secured on topics having direct bearing on U.S. security. Covert action should be resorted to only when the security of the United States is directly threatened, when statecraft can be shown not to work, and when the possible ill effects of the action do not outweigh its possible benefits. When issues are faced clearly, the authors of this book maintain, there will probably be no place in the post–Cold War world for covert operations, and a drastically limited role for espionage.

The Primacy of Public Diplomacy

Melvin A. Goodman, formerly with the CIA, seconds these notions. He states that CIA propaganda "has had little effect on foreign audiences and should end immediately," and that "Covert efforts to influence foreign elections or political parties should stop." The CIA's paramilitary activity should be transferred to the Defense Department, covert military interventions should be ended immediately, and a ban on covert action in peacetime altogether should be under discussion.

Goodman warns against any attempt to create a separate organization for espionage, as suggested by the House Intelligence Committee; such a move would lend automatic sanction to an activity that should be undertaken only when other methods cannot work. He also argues strongly against uniting the directorate of the CIA that carries out espionage and covert action, the Directorate of Operations, with the one that analyzes intelligence, the Directorate of Intelligence, as the Brown Commission suggested. This would give clandestine reports a privileged status that they don't deserve, and could unduly influence the analytic product of the CIA.

Goodman also states that it would be a step backward to endorse the expanded use of CIA cover to include journalists, clergy, and Peace Corps volunteers. Journalists, in particular, he believes, should be a check on government, not its instrument.

Like Hilsman and other contributors, Goodman believes that the interests of the United States would be better served by shifting resources otherwise devoted to clandestine collection to the U.S. Foreign Service, which has had to reduce its personnel in the field because of budget limitations. The result, Goodman and others argue, has made the United States unnecessarily dependent on the CIA, which has the money and personnel to carry on foreign relations and most intelligence collection, which should properly be in the hands of the diplomatic service.

Goodman maintains that intelligence should be the servant of policy, not its initiator. Given these considerations, and the desirability of a paramount role for public or open diplomacy, Goodman suggests that the CIA might well report to the State Department, rather than be maintained as a fully autonomous agency of government.

These ideas are underscored strongly in chapters by Robert E. White and Robert V. Keeley, former U.S. Ambassadors, who believe that the Foreign Service must be maintained as the premier agent of American foreign policy in the field, and that CIA personnel should be withdrawn altogether from friendly and many neutral areas. (Exceptions might be made for personnel necessary to maintain liaison with their intelligence counterparts from other countries.) White suggests that such arrangements might be reached by international agreement, and that the United States might well use Central America and the Caribbean as a test case.

Keeley spells out how the autonomous and increasingly aggressive role of the CIA in the field has worked to the disadvantage of the United

States. In addition to citing the problems of the CIA in relying on bought sources and the Agency's tendency to defend governments that it has supported, he recounts a number of cases where the CIA, at unnecessary cost, has attempted to insinuate itself as the agent of U.S. foreign policy in the field. Both Keeley and White recount ways in which the CIA, particularly, has engaged in destructive power struggles with the State Department in the field, subverting official U.S. policy, undermining the authority of the embassy, and withholding essential information from U.S. Ambassadors and policymakers. Both authors insist that responsible statecraft requires that the U.S. Ambassador represent the United States, and that in order to do so, he or she must be fully informed of all plans for clandestine actions well in advance of their possible execution.

THE PROBLEM OF BLOWBACK

The usurpation of U.S. foreign relations by the intelligence system, primarily by the CIA, has had harmful effects within this country as well as abroad. Jack A. Blum maintains that the image of an omnipresent CIA, acting as an extralegal body, is in itself damaging to democratic institutions. Instances of covert manipulation of domestic political affairs and subversion of law have fueled a popular belief in a ubiquitous CIA responsible for violations of citizens' rights. Although some of these fears are fictitious, their effect has been to undermine U.S. citizens' sense of the fairness and legality of their system of government.

Blum also points out the improper use of power by the CIA to block the President's choice of the Director of Central Intelligence, and maintains that CIA methods have encouraged foreign powers, such as Nationalist China, Korea, and India, to illegally attempt to influence U.S. foreign-policy legislation. He also argues that the intelligence community's "black budget" results in a lack of normal contracting controls, wasteful spending, and improper relationships with contractors in the absence of General Accounting Office audits and contract challenges—allegations supported by Robert Dreyfuss and Richard Stubbing. Blum suggests remedial legislation and administrative practices to curb such abuses.

REDUCING SECRECY

Kate Doyle shows how the intelligence community has repeatedly hidden behind official secrecy to remove its operations from legitimate public scrutiny, making difficult or impossible any reasoned assessment of important historical events. "We must demand that our Congress and our President broaden the incipient structural reforms that have been made to the secrecy and classification system to date, with the goal of finally dismantling the national-security state," Doyle writes.

To balance openness and the requirements of secrecy, Doyle proposes three innovations: the presumption of openness, a public-interest balancing test for secrecy, and the capacity for outside review. Specifically, she proposes that the presidential executive order on openness be extended to intelligence; if this is not forthcoming, Congress should set legislative limits on the intelligence community's ability to conceal information. A balancing test should also be part of the classification and declassification process, which should be subject to judicial review in deciding challenges to intelligence-agency denials under the Freedom of Information Act.

Doyle also recommends that the intelligence agencies initiate a historical-review process to release their oldest and most historically significant files to the National Archives, institute an annual declassification process, and reveal their aggregate budgets on a yearly basis. Accordingly, Congress should establish a clear and expeditious process whereby documents of public concern can be reviewed for declassification, and adequate funds should be provided to the FOIA offices so that they can do their work. In addition, Doyle suggests strengthening legislation and administrative procedures protecting U.S. citizens from infringements of their privacy by U.S. intelligence organizations.

PUTTING INTELLIGENCE OUT OF THE NARCOTICS BUSINESS

Perhaps no topic so illustrates the deleterious effects of covert operations as the long history of the CIA's support for international drug trafficking. Alfred W. McCoy richly documents the Agency's support of drug traffickers from Laos to Central America, and the protection of its sources from DEA prosecution. As McCoy makes clear, the CIA has left a domestic legacy of illegality, suspicion, and racial division. It has violated federal laws and later executive orders requiring it to provide intelligence in support of the drug war. Like the other authors of this book, McCoy insists that the CIA be held to standards of legality. "Now that the Cold War is over," he asks, "do we really need a CIA armed with the extraordinary powers that place it beyond the law?"

President Clinton recently assigned the CIA the task of "protecting American citizens from new transnational threats such as drug traffickers...." But, as McCoy, Blum, and others document, the CIA is hopelessly compromised by its past alliances. Accordingly, McCoy and others warn that the CIA should be prohibited from engaging in drug-policy enforcement, and that unless it is properly regulated, the CIA's tolerance of drug dealing is likely to recur.

Conclusions 217

REDUCING THE EXCESSES OF ELECTRONIC INTELLIGENCE

Robert Dreyfuss focuses on the heavily funded agencies involved in technical and electronic intelligence (TECHINT). These agencies play a vital role in monitoring military forces in threatening areas around the world, although, here again, the appropriate targets for the intelligence community have decreased with the end of the Cold War.

New tasks for TECHINT have been suggested in the areas of support for military operations, environmental intelligence, and economic espionage. But Dreyfuss and others raise serious questions about whether such intelligence can be justified "either in terms of costs and benefits or according to the traditional definition of national security." Regarding the use of TECHINT in support of the military, Dreyfuss states that budgetary decisions must be made on the basis of national-security needs and cost-benefit studies, not on the fact that the Department of Defense controls the budgets of the National Reconnaissance Office and the National Security Agency, a point that is emphasized throughout this book.

As for environmental and economic intelligence-gathering, Dreyfuss finds that the expensive hardware of the TECHINT agencies seems especially unsuited to such work. Like other authors in the book, such as Stubbing and Holt, he strongly believes that intelligence should not be used for economic espionage against foreign corporations. Technical intelligence can play an important role in monitoring nuclear proliferation and combating terrorism, but this role must be complementary to other means of securing information.

Dreyfuss points to another area of concern, one described by Pat Holt and others in this book as well: that of the compromising relations among intelligence contractors, the agencies, and congressional oversight committees. Such compromising relations, he and others maintain, should be prevented as much as possible by legislation and administrative practice.

FUNDING AND MANAGING THE INTELLIGENCE SYSTEM

The last two chapters in this book come from writers who have been concerned with monitoring the intelligence establishment within the U.S. government, Richard Stubbing through the Office of Management and Budget and Pat Holt through the Senate Foreign Relations Committee. The deficits of past intelligence operations and the changing picture of the world since the end of the Cold War, presented by the authors preceding them, prompt a reassessment of how intelligence should be managed and funded, and how it can be better supervised.

One difficulty, Richard A. Stubbing maintains, is that authority over the intelligence system is mixed and uncertain. Although the Director of Central Intelligence has statutory responsibility for foreign intelligence and the authority to participate fully in the preparation of the budget, he lacks the ability to impose his priorities on most of the agencies theoretically under him. Stubbing urges that the DCI should hire the directors of the NRO, the NSA, and the National Imagery and Mapping Agency (NIMA) in peacetime, and have primacy in operational assignments for these three agencies, in consultation with the Secretary of Defense. This should affect the undue amount of intelligence money going into military targeting. In the future, Stubbing writes, "Nonmilitary targeting should grow to account for 60 percent to 70 percent of the intelligence effort, with the remainder allocated to military needs." Control by the DCI, however, must occur only within a context of accountability, which must be established by the kinds of reforms suggested by the authors of this book.

Stubbing holds that consumer input, rather than bureaucratic leverage, should determine the allocation of intelligence resources and assignments. He therefore supports the formation of a Consumers Committee (proposed by the Brown Commission), which would have representation from all the principal producers and users of intelligence reports. "This group would report regularly to the [new NSC] Committee on Transnational Threats on how well the needs of intelligence consumers are being met, identifying gaps and shortcomings."

In the U.S. governmental system, the budget is usually the focal point of policy. But Stubbing finds the budget process for intelligence overwhelmed by detail and unable to deal with the basic issues. Agencies, he writes, "seldom take critical looks at such fundamental questions as the overall level of intelligence; the proper blend of imaging, electronic, and human intelligence collection; and needed improvements in intelligence analysis and output." He states that "For openers, a top-down review of requirements must replace the incremental budgeting approach used in the past."

Like other authors in this book, Stubbing suggests that more resources should go to the U.S. diplomatic service and fewer to CIA field operatives, and he agrees that intelligence resources should be aimed primarily at rogue or threatening states, terrorism, and the proliferation of nuclear weapons. The analytical capacity of the CIA should be enhanced, including the use of outside resources, and the analysts better trained.

Like Dreyfuss, Stubbing suggests that there should be less dependence on sophisticated satellite and ground systems, and that NSA voice and data intercepts should be reduced. He also believes that there has to be far better cooperation between the FBI and the CIA, and that the FBI needs strengthening in counterintelligence and counterterrorism. Following through on these and other suggestions in the book, he proposes budgetary cuts for the intelligence community over a five-year period that amount to $4.1 billion to $5.7 billion, or 15 percent to 20 percent.

Reform of the intelligence community, Stubbing believes, will require a DCI who is highly respected, has vision, is willing to set a new course, and is capable of withstanding criticism. The DCI will also require the full endorsement of the President and the congressional leadership. Steps to effect intelligence reform should include the establishment of a Presidential Commission on Future Threats and an NSC study of intelligence reform, to include representatives from all agencies involved with intelligence. After the President decides on a package for intelligence reform, legislative approval should be secured from Congress.

Pat M. Holt parlays his long experience as head of the Senate Foreign Relations Committee staff into a set of recommendations designed to strengthen public accountability and control of the intelligence community. How, he asks, can an open, democratic society such as the United States establish public control of activities that are necessarily secret?

Holt proposes that Congress tighten its legal provisions about its right to know and about the timeliness of the reports it receives on U.S. intelligence activities. Congress must no longer look upon itself as the advocate of intelligence, but rather as its supervisor. It must also look to various laws and procedures, such as the Classified Information Procedures Act, that enable the agencies to block legitimate inquiry into their activities. In addition, the Foreign Intelligence Surveillance Act needs to be strengthened to protect citizens' Fourth Amendment rights that are threatened by intelligence operations, particularly those of the NSA. Congress should also reconsider the idea of a full legislative charter regulating intelligence affairs.

Equally, the executive branch must strengthen its supervision of the Intelligence Committees. Holt recommends the strengthening of the President's Foreign Intelligence Advisory Board, which should be able to deal with interagency rivalries and command full access to insure results.

To curb the continuing ill effects of covert action, Holt believes, improved congressional surveillance is mandatory, because intelligence oversight is not generally susceptible to legislative solutions. Indeed, the difficulty of legislation is the most powerful argument for critical oversight. Up to now there has been a lack of political will in both the executive branch and Congress; this must change. Both the executive branch and Congress should subject proposed intelligence operations to more rigorous cost-benefit analysis, including an analysis of any short- and long-term risks involved in failure. Both the President and Congress must radically increase their surveillance of intelligence if the United States is to have a system that is both effective and reflective of democratic values. Holt recommends specifically that to increase public accountability, plausible deniability should be ruled out as a part of any intelligence operation. He also recommends that intelligence agencies be prohibited from receiving gifts or any external financial support that could influence an operation.

To make the congressional Intelligence Committees more reflective of the membership and less of the leadership, Holt believes, they should be standing, not select; party leaders should not have ex officio status; and service on the committees should not be subject to term limits. In addition, the Intelligence Committees should be given jurisdiction over all components of the intelligence community in order to end destructive rivalries with the Armed Services and Judiciary Committees.

Congress, Holt maintains, must require that the overall intelligence budget, and probably that of each agency, be made public, unless the agency can show cause for it to remain secret. Congress should review all intelligence legislation to eliminate reporting loopholes and protect civil liberties. Holt also proposes that Congress consider a ban on paramilitary covert action.

An Intelligence System for the Post–Cold War Era

The reforms outlined above project an intelligence system that moves away from the life-and-death struggle of the Cold War. It is a system that collects intelligence effectively but respects citizens' rights and international law.

What it moves toward, of course, is unknown. We do not know what kind of world we will have ten years from now, much less fifty or 100. But as the world's leading military and economic power, the United States will have a major influence. The way it designs its institutions and the standards it sets will indeed structure the future. It can thus work toward a world that differs materially from the violence-prone century from which we are emerging.

An intelligence system that subverts international and national law, gratuitously harms others, and corrupts political systems, as U.S. intelligence has, will be a detriment to building a better world. International agreements on the curtailment of espionage and covert action, such as Ambassador White proposes, will reverse century-old trends. As many of the writers of this book have said, many intelligence activities, and particularly their illicit aspects, are designed to fight other intelligence services. Curtail the activities, by mutual agreement or by tacit consent, and much of the illegality can be eliminated. As intelligence becomes more circumscribed in its dealings, the edgy aggressiveness of the world will be that much lessened.

These moves in the intelligence community can parallel those in the diplomatic sphere, as the United States moves to support the rule of international law in foreign relations. Rather than being one of the few nations of the world to oppose such measures as a convention on international criminal jurisdiction or on the regulation of land mines, the United States can begin to see its future in cooperative, multilateral terms, not in terms of the Cold War.

This is not to embrace a childish optimism. The country needs to move ahead step-by-step, watching its flanks and backside as it goes. There are likely to be, for some centuries, powerful and potentially unstable countries such as Russia, and rogue states such as Iraq and Sudan, that need to be closely watched; international terrorists who need to be stalked; nuclear materials that need to be tracked; and chemical and biological weapons that need to be monitored. An intelligence system capable of acquiring needed information on these developments will be part of the U.S. security machinery for some time to come. But rather than augmenting this kind of world, this country can work carefully to build a different one.

Nor need the capacities developed in the past fifty years be lost. U.S. intelligence resources can be pooled with the resources of other countries, as they will be in the new NATO arrangements, to fight international terrorism. They also hold possibilities for the monitoring of international agreements through the United Nations, if such arrangements are not used to gain clandestine information for the United States. The technology can also be put at the disposal of specialized agencies, such as the World Meteorological Organization and other international and national scientific organizations.

The writers of this book have outlined the first steps toward an intelligence system in keeping with a more law-abiding, cooperative world. Achieving such a system will require imagination and courage, and enormous perseverance by our President, key members of Congress, and the Director of Central Intelligence. But the task is worth undertaking if the United States is to have an intelligence system that both insures its security and conforms to its democratic vision.

Selected Bibliography

THIS BIBLIOGRAPHY is meant to provide a general introduction to the topic of U.S. intelligence, and a few of the salient references listed in the individual chapters. For a succinct description of the intelligence community, see Chapter 1 of Loch K. Johnson's *Secret Agencies*. For a more detailed description, including a discussion of military intelligence, see Jeffrey T. Richelson's *The U.S. Intelligence Community*.

BOOKS

Adams, James (1994). *The New Spies: Exploring the Frontiers of Espionage*. London: Hutchinson.
Ambrose, Stephen E. (1981). *Ike's Spies: Eisenhower and the Espionage Establishment*. Garden City: Doubleday.
Andrew, Christopher A. (1995). *For the President's Eyes Only: Secret Intelligence and the American Presidency from Washington to Bush*. New York: HarperCollins.
Bamford, James (1982). *The Puzzle Palace*. New York: Houghton Mifflin.
Brugioni, Dino (1990). *Eyeball to Eyeball: The Inside Story of the Cuban Missile Crisis*. New York: Random House.
Council on Foreign Relations (1996). *Making Intelligence Smarter: The Future of U.S. Intelligence*. New York: Council on Foreign Relations.
Day, Dwayne, John Logsdon, and Brian Latell, eds. (1998). *Eye in the Sky: The Story of the Corona Spy Satellites*. Washington, DC: Smithsonian Institution Press.
Duffie, Whitfield, and Susan Landau (1998). *Privacy on the Line: The Politics of Wiretapping and Encryption*. Cambridge, MA: MIT Press.
Gates, Robert M. (1996). *From the Shadows: The Ultimate Insider's Story of Five Presidents and How They Won the Cold War*. New York: Simon and Schuster.
Godson, Roy (1995). *Dirty Tricks or Trump Cards: U.S. Covert Action and Counterintelligence*. Washington, DC: Brassey's.
Godson, Roy, Ernest R. May, and Gary Schmitt, eds. (1995). *U.S. Intelligence at the Crossroads: Agendas for Reform*. Washington, DC: Brassey's.
Goodman, Allan E., Gregory F. Treverton, and Philip Zelikow (1996). *In from the Cold: The Report of the Twentieth Century Fund Task Force on the Future of U.S. Intelligence*. New York: Twentieth Century Fund Press.
Grose, Peter (1994). *Gentleman Spy: The Life of Allen Dulles*. New York: Houghton Mifflin.

Hilsman, Roger (1996). *The Cuban Missile Crisis*. Westport, CT: Greenwood Publishing Group.
——— (1999). *From Nuclear Military Strategy to a World Without War: A History and Proposal*. Westport, CT: Praeger.
Holt, Pat M. (1994). *Secret Intelligence and Public Policy: A Dilemma of Democracy*. Washington, DC: Congressional Quarterly Books.
Isacson, Adam (1997). *Altered States: Security and Demilitarization in Central America*. Washington, DC: Center for International Policy.
Jeffreys-Jones, Rhodri (1998). *The CIA & American Democracy*, 2nd ed. New Haven and London: Yale University Press.
Johnson, Loch K. (1996). *Secret Agencies: U.S. Intelligence in a Hostile World*. New Haven: Yale University Press.
Johnson, Robert H. (1994). *Improbable Dangers: U.S. Conceptions of Threat in the Cold War and After*. New York: St. Martin's Press.
Knott, Stephen F. (1996). *Secret and Sanctioned: Covert Operations and the American Presidency*. New York: Oxford University Press.
Lashmar, Paul (1996). *Spy Flights of the Cold War*. Annapolis: Naval Institute Press.
Leogrande, William M. (1998). *Our Own Backyard: The United States in Central America, 1977–1992*. Chapel Hill: University of North Carolina Press.
McClintock, Michael (1992). *Instruments of Statecraft: U.S. Guerrilla Warfare, Counterinsurgency, and Counterterrorism, 1940–1990*. New York: Pantheon Books.
McCoy, Alfred W. (1991). *The Politics of Heroin: CIA Complicity in the Global Drug Trade*. New York: Lawrence Hill Books.
Moynihan, Daniel Patrick (1998). *Secrecy*. New Haven and London: Yale University Press.
Olmsted, Kathryn S. (1996). *Challenging the Secret Government: The Post-Watergate Investigations of the CIA and FBI*. Chapel Hill: University of North Carolina Press.
Perry, Mark (1992). *Eclipses: The Last Days of the CIA*. New York: William Morrow.
Polmar, Norman, and Thomas B. Allen (1997). *The Spy Book: The Encyclopedia of Espionage*. New York: Random House.
Powers, Thomas (1979). *The Man Who Kept the Secrets: Richard Helms and the CIA*. New York: Alfred A. Knopf.
Prados, John (1996). *Presidents' Secret Wars: CIA and Pentagon Covert Operations from World War II Through the Persian Gulf*. Chicago: Elephant Paperbacks.
Ranelagh, John (1986). *The Agency: The Rise and Decline of the CIA*. New York: Simon and Schuster.
Reisman, W. Michael, and James E. Baker (1992). *Regulating Covert Action: Practices, Contexts, and Policies of Covert Coercion Abroad in International and American Law*. New Haven: Yale University Press.
Richelson, Jeffrey T. (1995). *The U.S. Intelligence Community*, 3rd ed. Boulder: Westview Press.
——— (1999). *America's Space Sentinels: DSP Satellites and National Security*. Lawrence: University Press of Kansas.

Richelson, Jeffrey T., and Desmond Ball (1985). *The Ties That Bind: Intelligence Cooperation Between the UKUSA Countries.* Boston: Allen and Unwin.
Riebling, Mark (1994). *Wedge: The Secret War Between the FBI and CIA.* New York: Alfred A. Knopf.
Rudgers, David F. (forthcoming). *Creating the Secret State: The Origins of the Central Intelligence Agency, 1943–1947.* Lawrence: University Press of Kansas.
Simpson, Christopher (1998). *Blowback.* New York: Weidenfeld & Nicolson.
Smist, Frank J. Jr. (1990). *Congress Oversees the United States Intelligence Community, 1947–1989.* Knoxville: University of Tennessee Press.
Stubbing, Richard A. (1986). *The Defense Game.* New York: Harper & Row.
Theoharis, Athan G., ed. (1998). *A Culture of Secrecy: The Government Versus the People's Right to Know.* Lawrence: University Press of Kansas.
Thomas, Evan (1995). *The Very Best Men: Four Who Dared—The Early Years of the CIA.* New York: Simon and Schuster.
Treverton, Gregory T. (1987). *Covert Action.* New York: Basic Books.
Weiner, Tim, David Johnston, and Neil A. Lewis (1995). *Betrayal: The Story of Aldrich Ames, an American Spy.* New York: Random House.
Woodward, Bob (1987). *Veil: The Secret Wars of the CIA, 1981–1987.* New York: Simon and Schuster.

ARTICLES

Dreyfuss, Robert (1995). "The CIA Crosses Over." *Mother Jones,* January/February.
——— (1995). "Spying on the Environment." *E: The Environmental Magazine,* February.
——— (1996). "Orbit of Influence: Spy Finance and the Black Budget." *American Project,* March–April.
Goodman, Melvin A. (1997). "Ending the CIA's Cold War Legacy." *Foreign Policy,* No. 106, Spring.
Hilsman, Roger (1995). "Does the CIA Still Have a Role?" *Foreign Affairs,* September/October.
Shane, Scott, and Tom Bowman (1995). "No Such Agency." *Baltimore Sun,* December 3–15.

GOVERNMENT DOCUMENTS (AVAILABLE THROUGH THE U.S. GOVERNMENT PRINTING OFFICE)

Commission on the Roles and Capabilities of the United States Intelligence Community (March 1, 1996). *Preparing for the 21st Century: An Appraisal of U.S. Intelligence.*
Permanent Select Committee on Intelligence, House of Representatives, One Hundred Fourth Congress (1996). *IC 21: The Intelligence Community in the 21st Century.*

About the Center for International Policy

THE CENTER FOR INTERNATIONAL POLICY was founded in 1975, in the wake of the Vietnam war, to promote a foreign policy based on democracy, social justice, and human rights. Using its own reports, the Internet, conferences, fact-finding delegations, and news-media interviews, the Center alerts members of Congress and American citizens to policies that undermine democratic values. The Center highlights the effects of U.S. security and economic-aid programs on developing countries, and the extent to which U.S. policies foster democracy and respect for human rights.

In the late 1970s the Center led a campaign to set human-rights standards for the distribution of U.S. foreign assistance. The cochairmen of the Center's board were Representative Donald Fraser and Representative (now Senator) Tom Harkin, the principal sponsors of landmark legislation that made the promotion of human rights a stated goal of U.S. foreign policy.

In the 1980s the Center's focus shifted to Central America. U.S. policy was fueling civil wars in El Salvador, Guatemala, and Nicaragua, and the Center led efforts in the United States to seek a negotiated regional solution. The Center's staff organized U.S. support for the Contadora and Arias peace plans, which culminated in an August 1987 agreement written by President Oscar Arias of Costa Rica and signed by all five Central American presidents.

Robert E. White, former U.S. Ambassador to El Salvador, who had taken a courageous stand against the death squads and in favor of a negotiated end to that nation's civil war, became the Center's president in 1989. When the Haitian military overthrew the democratically elected president of Haiti in 1991, Ambassador White took up the cause of Haiti's exiled democratic leaders. White led five congressional delegations to Haiti and wrote recommendations for the incoming Clinton administration. A Center team accompanied President Jean-Bertrand Aristide and other democratic leaders during the critical negotiations on Governors Island in 1993 that established the legal and diplomatic conditions for the restoration of the constitutional order in Haiti.

In 1992 the Center established a program to forge a more rational U.S. policy toward Cuba. Wayne S. Smith, formerly the top U.S. diplomat in Havana, joined the Center to head the Cuba project, which advocates an end to the U.S. trade embargo. The embargo, Smith argues, is counterproductive and hurts U.S. economic and political interests. The Cuba project holds conferences in the United States and Cuba, and facilitates exchanges between Americans and Cubans. A major goal of the project is to broaden the debate over U.S. policy toward Cuba. To this end, the Center has taken delegations of farm officials, health professionals, mayors, congressional staff members, and retired military officers to Cuba to meet with their counterparts on the island.

In 1993 the Center for International Policy and the Costa Rica–based Arias Foundation organized a joint project to encourage the demilitarization of Central America and the Caribbean. The project initiated a grassroots demilitarization campaign that now has chapters in all of the region's countries. The project has published articles and books, and in 1998, with the Latin American Working Group, copublished a book entitled *Just the Facts,* which for the first time documented all U.S. security programs in Latin America and the Caribbean.

In 1995 the Center began work on a project to reform U.S. intelligence agencies. Like the demilitarization project and those focusing on Haiti and Cuba, this project seeks to find democratic solutions to foreign-affairs issues facing the United States. The Center assembled an advisory council of academics, former senior government officials, and nongovernmental specialists to begin a thorough review of the activities of the CIA and other intelligence agencies. Over the following two years, the Center sponsored six seminars in congressional hearing rooms on intelligence issues. Ambassador White and Melvin Goodman, formerly a high-ranking CIA specialist on the Soviet Union, took the Center's reform campaign to cities in the South and West on a speaking tour sponsored by local chapters of the World Affairs Council.

In 1998, Craig Eisendrath, a senior fellow at the Center, began preparation of *National Insecurity: U.S. Intelligence After the Cold War.* The intelligence-reform project is drafting a National Intelligence Reform Act, based on the seminars and the book, which will be presented to the new President and Congress in 2001.

The Center is launching a national-security project that will expand the mandate of the intelligence-reform project to include Defense Department programs involved in the formulation and execution of U.S. foreign policy. The Center believes that misplaced funding priorities have contributed to the militarization of U.S. foreign policy and reduced public accountability.

Military and intelligence agencies' autonomy and initiative have grown to such an extent that they are in fact carrying out a parallel U.S. foreign policy in much of the world. As military contacts and priorities gain primacy over diplomacy, there is a much heavier reliance on military solutions to problems that could otherwise be solved through diplomatic channels. Just as the political process has become increasingly secretive and undemocratic, very often the policy itself undermines democratic institutions abroad.

The national-security project will begin by documenting the expansion of the military's role overseas. The project will publish regional databases on U.S. military activities, from arms sales to obscure military training programs. These books will follow the model of *Just the Facts,* the Center's analysis of defense programs in Latin America and the Caribbean. In each region of the world the Center will focus on countries that have seen a sharp increase in U.S. security assistance and a deterioration of citizens' human rights. In the Western hemisphere the Center is focusing on Colombia, which last year received the third-largest amount of worldwide U.S. security assistance. In the spring of 1999, Ambassador White led a congressional delegation to Colombia to meet with top government officials and with the leaders of the main guerrilla group, the FARC. It was the first time the guerrilla leaders had met with a member of the U.S. Congress. Center staff members are urging U.S. policymakers in the executive and legislative branches to support the Colombian government's efforts to negotiate a settlement with the guerrillas to end the thirty-year-old civil war that is destroying their country.

Two global campaigns to curb arms sales form the activist component of the national-security project. The Code of Conduct campaign would limit the supply of arms by banning the sale of weapons to countries with nondemocratic governments. The project tries to limit the demand for weapons with its Year 2000 campaign, which helps citizens' groups in less-developed countries press their governments to spend less on weapons and more on education and health care.

The Center for International Policy believes that military and intelligence agencies must be subjected to far greater scrutiny of their overseas activities to make them more accountable to the American people and their elected representatives. The resulting increase in transparency must be accompanied by a nongovernmental effort to spark a debate on U.S. security policy, a debate based on authoritative information and honest analysis. In a democracy, the broad direction of foreign and military policy should be decided in the first instance by the action of an informed citizenry.

About the Contributors

JACK A. BLUM, an international lawyer with Lobel, Novins and Lamont, was the chief investigator for the Senate Foreign Relations Committee, headed by Senator Frank Church, and for the Senate investigation of the Iran-Contra scandal.

KATE DOYLE is an analyst with the National Security Archive.

ROBERT DREYFUSS is a journalist who publishes regularly on intelligence matters.

CRAIG EISENDRATH (editor) is a senior fellow at the Center for International Policy and a former Foreign Service officer. For thirteen years he was executive director of the Pennsylvania arm of the National Endowment for the Humanities. He is also the cofounder of the National Constitution Center in Philadelphia.

MELVIN A. GOODMAN is former division chief and senior analyst at the CIA's Office of Soviet Affairs from 1976 to 1986. He is currently a professor of International Security Studies and chairman of the International Relations Department at the National War College.

ROGER HILSMAN is Professor Emeritus of Government and International Relations at Columbia University, and former Assistant Secretary of State for Far Eastern Affairs.

PAT M. HOLT is the former Chief of Staff of the Senate Foreign Relations Committee. He is the author of *Secret Intelligence and Public Policy,* published in 1994 by Congressional Quarterly Books.

ROBERT V. KEELEY was a Foreign Service officer for thirty-four years, serving as U.S. Ambassador to Greece, Zimbabwe, and Mauritius, and in numerous other postings in Europe, the Middle East, Africa, and Asia. After his retirement from the Foreign Service in 1989, he served as president of the Middle East Institute in Washington from 1990 to 1995.

ALFRED W. MCCOY is the author of *The Politics of Heroin: CIA Complicity in the Global Drug Trade,* published by Lawrence Hill Books in 1991. He teaches Southeast Asian History at the University of Wisconsin in Madison.

RICHARD A. STUBBING, currently on the faculty of the Terry Sanford Institute of Public Policy of Duke University, handled the intelligence budget for the U.S.

Office of Management and Budget for more than twenty years. He is the author of *The Defense Game,* published in 1986 by Harper & Row.

ROBERT E. WHITE served as U.S. Ambassador to Paraguay and El Salvador. He is currently the president of the Center for International Policy.

Index

Acheson, Dean, 53
Afghanistan: blowback from, 29, 76, 86–87, 89; CIA bias of intelligence on, 38–39; CIA involvement with drugs in, 119, 125–32, 139–41; CIA operations in, 29–30, 69, 77, 208; and Iran, 12; need for intelligence on, 12; and Pakistan, 33, 86–87, 127–32; as target of TECHINT, 156; U.S. intelligence reporting on, 178, 179
Aftergood, Steven, 41, 93, 114n
Algeria: need for intelligence on, 14; and terrorism by former Afghan resistance, 87
Allende, Salvador, 31, 32, 85, 194–95, 203
Alpirez, Julio Roberto, 29, 200, 204
Ames, Aldrich (Ames Affair), 18, 38, 77–78, 176
Angola: as target of TECHINT, 156; U.S. covert action in, 28, 32, 33, 198
Arafat, Yasir, 39
Arbenz Guzman, Jacobo, 19
Argentina: CIA activity in, 82, 88; and CIA and Contra war, 36; Cuban exiles in, 85; and the "disappeared," 77
Arias, Oscar, 58
Aristide, Jean-Bertrand, 53–54
Aspin, Les, 165
Augustine, Norman, 167–68

Begin, Menachem, 35
Belgium, economic spying by United States in, 164
Berlin blockade, 26
Bermudez, Enrique, 32, 134
blowback, 24, 29–30, 76–91, 212, 215; and Bay of Pigs veterans' activity, 76; and drugs in United States, 118–20, 130, 140–42; and erosion of U.S. democracy, 76–81; and World Trade Center bombing, 76

Bolivia: "cocaine coup" in, 88; Cuban exiles in, 85
Bosnia: Iranian arms sales to, 202; need for intelligence on, 182; as target of TECHINT, 156, 159, 166; weapons from CIA in, 29
Brazil, as target of TECHINT, 162, 164
Brezhnev, Leonid, 35
Britain, intelligence service of, 42
British Guiana, U.S. covert action in, 104
Brown, George, 152
Brown, Harold, 3, 177
Brown Commission (Commission on the Roles and Capabilities of the United States Intelligence Community), 3, 4, 34, 37–41, 176, 186, 214, 218; and TECHINT, 155, 157
budget of intelligence system, 1–3, 172–78, 196–97, 217–19
Bundy, McGeorge, 15
Burma, CIA involvement with drugs in, 119–23, 125, 126, 128, 139–42
Bush, George: and agreement with Boris Yeltsin to reduce missiles, 9–10; as badly informed by CIA, 26; as CIA director, 57, 191; and congressional oversight of intelligence, 202; and *Foreign Relations of the United States* series, 104; and Iran-Contra affair, 57, 193; and Iraqi invasion of Kuwait, 13, 83; and secrecy, 110

Cambodia, need for intelligence on, 11
Carter, Hodding III, 25
Carter, Jimmy: and Afghanistan, 127; and Ambassadors' authority, 65; and covert action, 24, 125; and El Salvador, 49, 50, 52; and intelligence budget, 177–78; and President's Foreign Intelligence Advisory Board, 192; and secrecy, 100–101, 110

Casey, William, 32–33, 38–39, 56–57, 82, 107, 163, 173, 175, 180
Castro, Fidel, 23, 30, 55–56, 77, 79, 84, 105
Center for International Policy, 227–30
Central America and Caribbean: CIA bias of intelligence on, 32, 38–39; CIA involvement with drugs in, 120, 133–42, 216; CIA operations in, 28, 33, 39–40, 45–60, 79, 82; Cuban exiles in, 85; and Esquipulas II peace treaty, 58; need for diminished role of CIA in, 36, 59, 214; release of U.S. documents on, 111; U.S. use of criminals in, 36, 88
Central Intelligence Agency Act, 109
Chacon, Juan, 51–52
Chamorro, Violetta, 34
Chile: Cuban exiles in, 85; release of U.S. documents on, 112, 203; U.S. covert action against, 31–32, 194–95, 200
China, communist: CIA operations in, 28; as focus of U.S. intelligence effort in Cold War, 176, 180, 181; leak of U.S. nuclear secrets to, 176; military budget of, 181; need for intelligence on, 10, 36, 182, 213; nuclear-missile capability of, 15; release of secret files by, 101; and Republic of China, 11; sale of military technology by, 14; as supplier of arms to Afghan resistance, 127; as supplier of missiles to Iran and Pakistan, 14; as target of TECHINT, 158; and Tiananmen Square, 28, 179; and U.S. bombing of embassy in Belgrade, 3; U.S. intelligence reporting on, 36, 179, 181, 182
China, Republic of (Nationalist China): CIA involvement with drugs in, 119, 121–22, 140, 143; use of U.S. covert techniques in United States by, 80, 81, 215
Christopher, Warren, 50
Church, Frank, 24, 40, 54, 195, 209
Church Committee, 9, 24, 39–40, 55, 76, 100, 195, 206
CIA (Central Intelligence Agency): assessment in Brown Commission report, 3, 4; assessment by Arlen Specter, 2–3; assessment by George J. Tenet, 2; assessment by President Truman, 21; and blowback, 76–91; budgetary control of, 172–89, 218–19; in Central America and Caribbean, 45–60; and CIA Information Act, 108–9; conduct of covert action and espionage by, 23–44, 213–14, 162; and connections with contractors, 165, 167; creation of, 21, 23, 26, 53, 193; destruction of records by, 3; double agents in, 3, 18, 38, 77–78, 176; and drugs, 118–48, 216; and economic intelligence, 163–64; and Foreign Service, 26, 27, 45–60, 61–75, 81, 214–15; leadership of, 172; militarization of, 41; oversight of, 190–211; proposed merger of DO and DI in, 37–38, 214; proposed reduction in role of, 21–22, 218; and secrecy, 92–117; and TECHINT, 149, 151–53, 156, 186–87. *See also* covert action
Classified Information Procedures Act, 206–7, 211, 219
Clinton, Bill: as badly informed by CIA, 26, 40; and CIA and drugs, 143, 216; and conspiracy stories, 79; and covert action, 24; and Haiti, 53–54; and Intelligence Oversight Board, 47, 111–12; and President's Foreign Intelligence Advisory Board, 192; and secrecy, 106–12; as supported by governments of China and Indonesia, 81; on twenty-first–century threats, 16

Cohen, Warren, 103
Colby, William, 96, 200
Cold War: in Africa, 69; covert action during, 19–20, 77; and drugs, 118–20, 129, 140, 142–44; end of, as basis of change of U.S. intelligence policy, 2, 22, 33, 40, 41, 76, 111, 150, 181, 212–13, 220–21; failure of U.S. intelligence during, 9, 78–80; as focus of TECHINT, 149, 156, 158, 163, 177, 183; as focus of U.S. intelligence effort, 1–2, 8, 172–74, 185; secrecy during, 92–93, 97, 101–2, 106
Columbia, and drugs, 133, 136–37
Combest, Larry, 3, 40
COMINT law, 99
the Congo (Zaire): assassination attempted by United States in, 31; CIA operations in, 69; U.S. covert action in, 104, U.S. support of Mobutu in, 33
Costa Rica: and drugs, 133; and Institute of Political Education, 194

Council on Foreign Relations, 36
counterintelligence, 82, 176; and citizens' rights, 206; during Cold War, 8; overseas, 18, 66, 72–73; and Venona Project, 107–8
covert action, 1, 8, 19–22, 23–34, 183, 190–91; Ambassadors' control of, 65; anecdotes of, 66–73; as assassination, 31, 40, 45, 55, 77, 79, 143, 208; authority for, 26; blowback effects of, 76–91; budgetary control of, 172–89; in Central America and Caribbean, 19, 45–60, 120; and criminals, 143–44; and drugs, 118–48; as election rigging, 1; oversight of, 190–211; as paramilitary force, 1, 41, 74, 208, 211, 214, 220; as political subversion, 1, 40, 19; reassessment of, 21–22, 25–27, 213–14; reciprocal agreements to limit, 59; and secrecy, 92–117. *See also* CIA; *individual country listings*
criminals, in U.S. covert action, 88–89, 143–44, 186
cryptography, 17, 22, 99, 154–55, 169. *See also* National Security Agency
Cuba: assassination attempted by U.S. in, 31, 55, 77, 79; and Bay of Pigs, 20, 25, 30, 32, 41, 55–56, 84, 99, 108, 109, 192, 194; blowback from U.S. operations in, 76, 84–86, 89; declassification of material on, 105; espionage against United States by, 35, 55; and Che Guevara, 28; and JM/WAVE, 55–56, 84–86; and nuclear-missile crisis, 10, 108, 178; as target of TECHINT, 150; U.S. intelligence reporting on, 35, 55, 178
Cyprus, 72–73
Czechoslovakia: coup d'état in, 26; openness in, 77; and Soviets, 35, 178–79

D'Aubuisson, Roberto, 50
Defense Department (Pentagon): budget of, 150, 174, 180, 188, 196, 197; and Central America, 53; and CIA, 26, 30, 40; and congressional oversight of intelligence, 189, 190–91, 203, 207–8; creation of, 23; and environmental intelligence, 161–63; and *Foreign Relations of the United States* series, 102; and intelligence budget, 150, 174, 175, 177; and

Iranian hostages, 94; and paramilitary capability, 41; and release of documents on El Salvador, 107; and secrecy, 94, 102; and TECHINT, 37, 150–52, 156, 159–60, 161–62, 165–66, 179, 180, 217
Defense Intelligence Agency, 46, 110, 153; budget of, 173–75, 184, 188; oversight of, 201
Deutch, John, 7n, 38, 106, 109, 118, 135, 162, 165–66, 201, 205
Devine, Michael, 29, 200
Didion, Joan, 55–56
Dominican Republic: assassination attempted by U.S. in, 31; U.S. invasion of, 194
Doolittle, Jimmy, 27
Drug Enforcement Administration (DEA): and Central America, 133, 138; and CIA, 32–33, 134, 139, 141; and Pakistan, 130–31
drugs, 118–48, 186, 216; in Afghanistan, 87, 125; in Bolivia, 88; in Burma, 121–22; in Central America, 118–20, 125; and CIA involvement to support Contras, 32–33, 91n; in covert action, 89; in diplomatic reporting, 63; in Latin America, 36; in Pakistan, 33; in South Africa, 33; and TECHINT, 150. *See also* Drug Enforcement Administration
Duarte, Napoleon, 51
Dulles, Allen W., 1, 19
Dulles, John Foster, 26–27

Eastern Europe: and Cold War, 27, 142, 176, 178, 181; need for intelligence on, 10; U.S. intelligence reporting on, 35
Eban, Abba, 59
economic espionage, 4, 23, 40, 150, 159, 163–64, 169, 182
economic reporting, 1, 179, 182
Egypt: as supplier of arms to Afghan resistance, 127; and terrorism by former Afghan resistance, 30, 87; U.S. intelligence reporting on, 35, 179
Eisenhower, Dwight D., 94, 204; and covert action, 24, 26; and defense budget, 26; and President's Foreign Intelligence Advisory Board, 192; and U-2 affair, 192

El Salvador, 49–53; and Nicaraguan Contra war, 32–33, 56; release of U.S. documents on, 107, 111; and U.N. Truth Commission, 111; U.S. covert action in, 28, 46, 82; and U.S.-trained Honduran commandos, 31; U.S. training of security forces in, 29, 45
environmental intelligence, 150, 159, 160–63, 169, 182; as function of TECHINT, 156
espionage, 18–19, 34–37; in Central America, 45–60; reciprocal agreements to limit, 59; by spies posing as diplomats, 62–63; in support of CIA covert action, 64. *See also* CIA; Cold War; *individual country listings*
Ethiopia: CIA operations in, 31, 69–70; as target of TECHINT, 156

FBI (Federal Bureau of Investigation): budget of, 173, 175, 183, 185, 188; and counterintelligence and counterterrorism, 164, 176, 218; and domestic espionage, 101, 162; and drugs, 143; and Greece, 73; and intelligence reform, 189; oversight of, 201, 208; and secrecy, 105, 110; surveillance of Martin Luther King by, 78–79; and TECHINT, 151, 154, 158
Federal Records Act, 95
Fiers, Alan, 57, 137, 138
Ford, Gerald: and Ambassadors' authority, 75n; and congressional oversight of intelligence, 202; and investigation of intelligence violations of citizens' rights, 195
Foreign Intelligence Surveillance Act, 206, 219
Foreign Relations of the United States (FRUS), 101–4, 116n
Foster, Vincent, 79
France: CIA involvement with drugs in, 119; and collaboration with Nazis, 77; as drug producer, 124; drugs and covert action in Vietnam by, 122; economic spying by United States in, 24, 40, 164; and terrorism by former Afghan resistance, 87; U.S. covert action in, 23; U.S. use of criminals in, 88

Freedom of Information Act (FOIA), 92–93, 96–97, 100, 105, 108, 109, 111, 113, 216

Gates, Robert M., 24, 25, 26, 29, 33, 35, 38–39, 93, 96, 107, 108, 137, 161, 180
Gephardt, Richard A., 201
Germany (East and West): and declassification of U.S. materials on Nazi war crimes, 106, 108; economic spying by United States in, 24, 40; and Nazi and Stasi past, 77; release of secret files by, 101
Gingrich, Newt, 167–68, 201
Goldwater, Barry, 205–6
Gorbachev, Mikhail, 19
Gore, Albert, 160–61
Goss, Porter J., 2, 39, 205
Greece, CIA operations in, 70–73
Guatemala: CIA involvement with drugs in, 119; overthrow of Arbenz government in, 19, 45, 99; release of U.S. documents on, 109, 111–12, 203; secrecy of U.S. operations in, 94–95, 200–201, 209; U.S. covert action in, 29, 80, 203, 204; U.S. intelligence reporting on, 3
Guatemalan Commission on Historical Verification, 3, 46
Guevara, Che, 28

Haig, Alexander, 52
Haiti, 53–55; CIA involvement with drugs in, 119; as target of TECHINT, 160; U.S. covert action in, 29, 53–55, 82; U.S. intelligence reporting on, 179, 182; U.S. training of security forces in, 29
Hall, Keith, 157, 205
Haq, Fazle, 128–30
Harrington, Michael, 200, 203
Hekmatyar, Gulbuddin, 30, 127–29, 132, 139
Helms, Jesse, 49, 53, 106
Helms, Richard, 32, 126, 194
Henderson, Loy, 103
Hillenkoetter, Roscoe H., 26
Hitz, Frederick P. (Hitz report on CIA and drugs), 135–36, 139
Honduras, 48–49; and drugs, 133, 136–39; and Nicaraguan Contra war, 57, 133, 136; U.S. covert action and

training of security forces in, 29, 31, 33, 45, 46, 88
House Permanent Select Committee on Intelligence, 39, 195–211; advocacy of intelligence agencies by, 42; and Armed Services Committees, 195–211; and CIA and drug trafficking, 89; and connections with intelligence community and corporations, 164–69, 204–6, 220; on covert action, 25; creation of, 190, 195; and intelligence budget, 2, 37, 174; and intelligence reform, 188–89, 190; and journalistic cover for intelligence operations, 36; and militarization of intelligence, 41; report of, 3, 36, 40; support for separate espionage organization by, 34, 214; and TECHINT, 150, 152, 155, 157, 160, 166–69, 204–6; and George J. Tenet, 37; and whistle-blowers, 95
HUMINT (human intelligence), 35, 40, 63
Hungary: and Soviets, 35, 178; U.S. covert action in, 32
Hussein, Saddam, 1, 15, 31, 33, 35, 39
Hyde, Alan, 136–38

India: as drug producer, 121, 125; economic spying by United States in, 40; need for intelligence on, 11–12; and nuclear tests, 1, 11–12, 24, 37, 158–59, 175, 179, 182; as target of TECHINT, 158–59; use of U.S. covert techniques in United States by, 81, 215; U.S. intelligence reporting on, 158–59, 175, 179, 182
Indonesia: U.S. covert action in, 31, 104; use of U.S. covert techniques in United States by, 81
Information Security Oversight Office, 94, 106
Inman, Bobby, 38, 177, 192
Inouye, Daniel, 101
Intelligence Identities Protection Act, 207
Intelligence Oversight Board, 3, 47, 111–12, 201
Iran: and Afghanistan, 12; arms sales to Bosnia by, 202; CIA cooperation with, 28; destruction of CIA files on, 96; as drug producer, 125–26; and *Foreign Relations of the United States* series, 102–3; military budget of, 181; need for intelligence on, 12, 14, 213; nuclear-missile capability of, 13–14, 15, 158; and nuclear purchases from China and Russia, 14; overthrow of Prime Minister Mossadegh in, 3, 19, 28, 102–3; overthrow of Shah in, 28, 127, 179, 209; as target of TECHINT, 156, 158; U.S. intelligence reporting on, 179, 180, 209
Iran-Contra scandal, 23–24, 25, 40, 57, 79, 95, 101, 192–93, 198, 207, 209
Iran-Iraq war, 12, 15, 178, 179
Iraq: and bank-loan scandal, 82–83; chemical and biological capability of, 16; military budget of, 181; need for intelligence on, 12–13, 213, 221; nuclear-missile capability of, 12–13, 14–15, 34; rescue of U.S. diplomats in, 28; as rogue state, 4, 39, 221; as target of TECHINT, 158; and tensions with other Middle Eastern nations, 12–13; U.N. inspection of, 3, 12–13, 33; U.S. covert action against, 1, 3, 31, 33; and U.S. covert relations with Kurds, 32; U.S. intelligence reporting on, 3, 35, 179
Israel: and CIA monitoring of peace accords, 39; Egyptian invasion of, 179; need for intelligence on, 12, 14; nuclear-missile capability of, 12, 14; and tensions with other Middle Eastern nations, 12, 14; U.S. intelligence reporting on, 35, 179
Italy: economic spying by United States in, 24, 40; and Pakistani drugs, 132; U.S. covert action in, 23, 80, 98; U.S. use of criminals in, 88

Japan: attack on Pearl Harbor by, 61, 97; as center for gathering intelligence, 176; economic spying by United States against, 24, 40, 164; and U.S. breaking of wartime codes, 17; U.S. covert action in, 104; U.S. use of criminals in, 88
John Paul II, 180
Johnson, Lyndon B.: and congressional oversight of intelligence, 202; and covert action, 24; as informed by CIA, 78; and invasion of Dominican Republic, 194
Joint Military Intelligence Program (JMIP), 173–74, 183, 184, 188
Justice Department: and Iraq bank scandal, 82–83; and oversight of intelli-

gence, 32, 200–201, 207; protection of drug dealers by, 134; and whistle-blowers, 95

Kennan, George, 26
Kennedy, John F.: and Ambassadors' authority, 65; assassination of, 78–79, 104–5; and Bay of Pigs, 23, 84, 194; and congressional oversight of intelligence, 202; as informed by CIA, 78; and National Reconnaissance Office, 98; and President John F. Kennedy Assassination Records Collection Act, 104–5, 108; and President's Foreign Intelligence Advisory Board, 192
Kenya: bombing of U.S. embassy in, 87, 208; U.S. intelligence reporting on, 179, 182
Kerry, John, 80, 133–34, 136
Khomeini, Ayatollah, 15
Khrushchev, Nikita, 35, 91n
Kim Il Sung, 11
Kim Jong Il, 11
King, Martin Luther, 78–79
Kissinger, Henry, 31–32, 70, 180
Korea, North: military budget of, 181; need for intelligence on, 11, 213; nuclear-missile capability of, 1, 11, 14, 34, 158; as target of TECHINT, 158; U.S. intelligence reporting on, 3, 179
Korea, South: Chinese invasion of, 179; and North Korea, 11; U.S. covert action in, 104; use of U.S. covert techniques in United States by, 81, 215
Kosovo: bombing of Chinese embassy in, 3; need for intelligence on, 182; and Russia, 10
Kuwait, Iraqi invasion of, 35, 83, 179

bin Laden, Osman, 30
Laos: CIA involvement with drugs in, 119–20, 122–24, 128, 134, 139–42, 216; U.S. covert action in, 20, 28, 32
Lebanon: Menachem Begin's intentions toward, 35; CIA training of terrorists in, 31; Israeli invasion of, 179
Lee Kuan Yew, 18
Letelier, Orlando, 85
Libya: bombing of Berlin nightclub by, 199; military budget of, 181; as target of TECHINT, 158; U.S. intelligence reporting on, 13
Lilley, James, 28

MacFarlane, Robert, 57
the Mafia, 88, 132
Magsaysay, Ramon, 20
Marshall, George C., 23
Marshall, Thurgood, 109
McAfee, Marilyn, 47
McCarthy, Joseph R., 202–3
Mengistu, Haile-Maryam, 31
Mexico: CIA involvement with drugs in, 119; and money laundering by Richard Nixon, 80; release of U.S. documents on, 112
militarization of intelligence, 229
Mobutu Sese Seko, 33, 69
Mossadegh, Muhammad (Mosadeq), 3, 19, 102–3
Moynihan, Daniel Patrick, 45, 106, 115n, 206
Moynihan Commission, 93, 94, 106
Mozambique, 28
Mullah Nasim Akhundzada, 131–32
Muskie, Edmund, 80

National Foreign Intelligence Program (NFIP), 173–74, 183, 188, 196
National Imagery and Mapping Agency (NIMA), 37, 149, 152–55, 157, 179; budget of, 169n, 173–75; DCI's control of, 186–87, 218; oversight of, 201
National Reconnaissance Office (NRO), 114, 149–71; budget of, 82, 109, 160, 169n, 173–75, 177, 183, 188, 217; and corporate connections, 164–68; creation of, 98, 149; DCI's control of, 186–87, 218; financial mismanagement of, 41, 82, 150; secrecy about, 107, 115n
National Security Act of 1947, 23, 97, 99, 113, 190
National Security Agency, 38, 149–71, 192; budget of, 109, 169n, 173–77, 183–84, 188, 217; and citizens' rights, 219; and corporate connections, 149; creation of, 98–99, 149; and cryptography, 17, 22; DCI's control of, 186–87, 218; and Defense Department, 160; financial mismanagement of, 150; and

Index   239

National Security Agency Act, 99; oversight of, 201; and secrecy, 99, 107
National Security Council (NSC): and covert action, 37, 42, 81, 104; creation of, 23, 97–98; and Iran-Contra affair, 193; role of, in intelligence, 31, 32, 79, 80, 97–99, 186–87, 189, 190–91, 218–19; and TECHINT, 156
Nicaragua: and CIA Inspector General's Report on the Contras and Drugs, 91n; CIA involvement with drugs in, 118, 125, 133–39, 141, 143, 144; and CIA manual for assassination, 31; and congressional action to end Contra war, 197; and Cuban exiles, 85; and Reagan administration's campaign in United States in support of Contra war, 81; secrecy of U.S. covert action in, 193; as target of TECHINT, 156; U.S. covert action in, 25, 28, 32–33, 34, 46, 56–57, 80, 206; and U.S. training of security forces, 29, 45
Nicholson, Harold J., 3
Nixon, Richard M.: and Chile, 195, 203; and congressional oversight of intelligence, 202; and covert action, 24, 80, 91n; and executive order on classification, 100; and Haile Selassie, 70; and war on drugs, 124–25, 143; and Watergate scandal, 79, 95, 100
Nordeen, Bill, 73
Noriega, Manuel, 36, 54
North, Oliver, 57, 81, 88
Novo Sampol brothers, 84–85, 91n
Nuccio, Richard, 94–95, 200–201

Office of Strategic Services (OSS), 19, 61
Oswald, Lee Harvey, 105
oversight of intelligence, 41–42, 190–211; by Congress, 193–211, 219–20; by executive branch, 174–75, 191–93, 210–11; and TECHINT, 164–68. *See also* House Permanent Select Committee on Intelligence; President's Foreign Intelligence Advisory Board; Senate Select Committee on Intelligence; *individual listings of U.S. Presidents*

Pahlavi, Mohammad Reza (Shah), 28, 102–3, 126, 179, 209

Pakistan: and CIA involvement in Afghanistan, 33, 86–87, 127–32; CIA involvement with drugs in, 119, 139–40, 142; as drug producer, 125–32; need for intelligence on, 11–12; and nuclear purchases from China, 14; nuclear-weapons program of, 11, 33, 182; as target of TECHINT, 158; and terrorism by former Afghan resistance, 30
Panama: and Noriega and Contra war, 36; and recruitment of Noriega, 54; U.S. training of security forces in, 45
Papandreou, Andreas, 71–72
Papandreou, George, 71
Peace Corps, and CIA, 36, 68, 203, 214
Perry, William, 165–66
Peru, terrorism in, 28
Philippines, and U.S. covert support for Magsaysay, 20
Phoumi Nosavan, 20
Pike, John, 151, 153, 156, 158
Pike, Otis J. (Pike Committee), 24, 39–40, 195
Pinochet, Augusto, 85, 112
Poindexter, John, 57, 193
Poland: intelligence service of, 28; U.S. covert action in, 28, 80
Powers, Gary, 17
President John F. Kennedy Assassination Records Collection Act, 104–5
Presidents (U.S.): and CIA, 65; and need for intelligence, 8–9; oversight of intelligence by, 191–93. *See also individual listings*
President's Foreign Intelligence Advisory Board (PFIAB), 192, 219
proliferation of weapons, 63, 74, 90, 175, 184–86, 213, 221; bacteriological, 2, 41, 182, 183, 185; chemical, 2, 41, 182, 185; as fostered by illegal trade, 4, 11; in Iraq, 34; in North Korea, 34; nuclear, 2, 36–37, 41, 182–84, 218; as target of TECHINT, 183

Reagan, Ronald: as badly informed by CIA, 26; and domestic covert action, 81; and El Salvador, 52; and intelligence budget, 173; and Iran-Contra scandal, 23–24, 56–57, 79, 95–96, 192–93, 198; and Pakistan, 128; and President's

Foreign Intelligence Advisory Board, 192; and secrecy, 100–101, 105, 106, 107, 110; and TECHINT, 173
Romero, Oscar Arnulfo, 50, 51
Roosevelt, Franklin D., 97
Roosevelt, Kermit, 102
Rumsfeld, Donald H., 14–15

Sadat, Anwar, 30, 35
Saltonstall, Leverett, 193
Saudi Arabia: and Afghanistan, 86–87, 127; and Egypt, 12; and Iraq, 12; and terrorism by former Afghan resistance, 30; U.S. intelligence reporting on, 179, 182
Schwarzkopf, H. Norman, 13, 179
secrecy, 9, 21, 38, 92–117, 198–203; and budgets for classification/declassification, 172, 185, 188, 216; CIA cult of, 63, 82; and covert action, 77, 90; and declassification of information, 111; and *Foreign Relations of the United States* series, 102–4; and Interagency Security Classification Appeals Panel (ISCAP), 110; and MKULTRA Program, 95; and plausible deniability, 21, 30, 82, 192–94, 210, 220; and "sources and methods" defense, 65, 76, 108, 113, 199. *See also* Freedom of Information Act; Moynihan Commission; President John F. Kennedy Assassination Records Collection Act
Selassie, Haile, 69–70
Senate Select Committee on Intelligence, 33, 34, 195–211; advocacy of intelligence agencies by, 42; and Armed Services Committees, 195–211; and CIA and drug trafficking, 89; and connections with intelligence community and corporations, 164–69, 204–6, 220; creation of, 190, 195; and intelligence budget, 3, 37, 174; and intelligence reform, 188–89, 190; and TECHINT, 150, 166, 169, 204–6; and George J. Tenet, 37
Shultz, George, 180, 193
Singapore, 18
Smith, Walter Bedell, 20
Smith, William French, 135
Somalia: as target of TECHINT, 156, 160; U.S. intelligence reporting on, 179, 182

Somoza, Anastasio, 56, 136
South Africa: Truth and Reconciliation Commission in, 77; U.S. covert action in, 33; use of U.S. covert techniques in United States by, 80–81;
Soviet Union (Russia), 178–81; and Afghanistan, 29, 86–87, 125, 127–32; after the Cold War, 4, 10, 142; and the Congo, 69; control of information in, 90n; declassification of material on, 105; espionage against United States by, 18–19, 26, 40, 66, 77–78, 107; and Ethiopia, 70; as focus of U.S. intelligence and covert action, 2, 8–9, 14, 22, 24, 25, 27, 33, 36, 41, 65–66, 97, 98, 99, 107–8, 149–50, 158, 175, 176, 178, 181, 213, 221; intelligence service of, 28, 154, 206; military budget of, 181; and nationalities problem, 10; nuclear-missile capability of, 10, 15; release of secret files by, 101; and sale of nuclear technology to Iran, 14; as target of TECHINT, 149–50, 156, 158, 161, 163, 177–78; U.S. intelligence reporting on, 1, 9, 17, 18–19, 26, 35, 38, 172, 178–82, 209; U.S. use of criminals in, 88
Specter, Arlen, 2–3, 202
Stalin, Joseph, 91n, 102
State Department (Foreign Service): budget of, 35, 50, 59, 66, 74, 75n, 172–73, 175, 184, 188, 214, 218; and CIA, 26, 27, 45–60, 61–75, 90, 214–15; and congressional oversight of intelligence, 203; and drugs, 131; and *Foreign Relations of the United States* series, 102–4; and intelligence reform, 189; and national security, 190–91; need for greater role of, 35, 36, 50, 58, 183, 214; and Nicaragua, 133; and Pakistan, 127, 130–31; and release of documents on El Salvador, 107; and TECHINT, 151; and whistle-blowers, 200–201
Stimson, Henry L., 17
Sudan, 4, 24, 29, 36, 200, 221
Sukarno, 31
Syria, military budget of, 181

Tactical Intelligence and Related Activities (TIARA): budget of, 173–74, 188; congressional oversight of, 196

Tanzania: bombing of U.S. embassy in, 87, 208; U.S. intelligence reporting on, 179, 182
TECHINT (technical intelligence), 1, 8, 17, 107, 149–71, 210, 217; corporate connections with, 152, 164–69, 204–6, 215. *See also individual country listings*
Tenet, George J., 2, 34, 37, 109, 173, 197, 205
terrorism: bacteriological and chemical, 16; in Greece, 73; in Japan, 16; nuclear, 15; in Saudi Arabia, 179; as target of U.S. intelligence, 36, 37, 41, 63, 74, 77, 90, 150, 175, 182–86, 213, 218, 221; against U.S. targets, 2, 16, 76, 87, 179, 208; by U.S.-trained Afghan resistance, 30, 87
Thailand, CIA involvement with drugs in, 119, 121–23, 143
Torricelli, Robert, 80, 95, 200–201, 203
Truman, Harry S.: assessment of CIA by, 21, 42; and congressional oversight of intelligence, 202; and establishment of intelligence system, 97, 208; and invasion of southwestern China, 121–22; and the press, 98–99
Turkey: and Cyprus, 72; as drug producer, 119, 121, 124, 126
Turner, Stansfield, 177, 197

United States Information Agency (USIA), 35, 59
U-2 spy planes, 99, 192

Vance, Cyrus, 34
Velasquez, Efrain Bamaca, 29, 200
Venezuela, Cuban exiles in, 85
Vietnam (North and South): assassination attempted by U.S. in, 31; Chinese invasion of, 179; CIA involvement with drugs in, 122–24, 140–42; declassification of material on, 95, 105–6; ending of war in, 197; need for intelligence on, 11; as target of TECHINT, 156; and tensions with China and Cambodia, 11; and Tonkin Gulf incident, 202; U.S. covert action in, 28, 77; U.S. debate over, 83, 194; U.S. intelligence reporting on, 179
Villeda Morales, Ramon, 48

Waters, Maxine, 118, 134
Webster, William, 26
Weinberger, Caspar, 57, 166, 193
Welch, Richard, 73
Woolsey, James, 26, 38, 53, 108, 109, 158, 164, 167

Zepeda, Thomas, 133
Zia ul-Haq, Mohammad, 127–31